HISTORY OF IDEAS SERIES

No. 6

THE HISTORY OF IDEAS SERIES

Under the editorial sponsorship and direction of the Editorial Committee, *The Journal of the History of Ideas*: John Herman Randall, Jr., chairman, George Boas, Gilbert Chinard, Paul O. Kristeller, Arthur O. Lovejoy, Marjorie H. Nicolson, Philip P. Wiener.

1. ROUSSEAU-KANT-GOETHE: *Two Essays*. By Ernst Cassirer, translated from the German by James Gutman, Paul Oskar Kristeller, and John Herman Randall, Jr.

2. NEWTON DEMANDS THE MUSE: *Newton's "Opticks" and the Eighteenth Century Poets*. By Marjorie H. Nicolson

3. THE IDEA OF USURY: *From Tribal Brotherhood to Universal Otherhood*. By Benjamin N. Nelson

4. TOLSTOY AND CHINA. By Derk Bodde, with the collaboration of Galia Speshneff Bodde

5. MORE'S UTOPIA: *The Biography of an Idea*. By J. H. Hexter

6. DIDEROT AND DESCARTES: *A Study of Scientific Naturalism in the Enlightenment*. By Aram Vartanian

DIDEROT AND DESCARTES

A Study of Scientific Naturalism in the Enlightenment

BY ARAM VARTANIAN

PRINCETON, NEW JERSEY
PRINCETON UNIVERSITY PRESS
1953

London: Geoffrey Cumberlege, Oxford University Press

L. C. CARD 52-8781

Printed in the United States of America
by Princeton University Press at Princeton, New Jersey

PREFACE

The history of ideas deals primarily with ideas that have, in the proper sense, a *history*. Perhaps the εἶδος does not quite enjoy that changeless and absolute status which Plato ascribed to it outside time and space; but there are, nonetheless, ideas whose successive development, or life cycle, may be viewed as in itself constituting a form of organic existence in relation to which places, periods, and persons, however important or significant, remain subordinate. The "historical transcendence" of certain ideas must often be a tacit assumption on the part of their biographer.

This book is concerned with such an idea: namely, that matter and its inherent modes of behavior have brought about all things, or at least (for the restriction is pertinent to our study), that all things are to be explained by recourse to matter and its properties. The career of this notion, since its rise at the dawn of classical antiquity, has almost wholly consisted in progressively more exacting definitions of matter and in more rigorously verifiable theories about its operations. In that still continuing history, we shall here focus our attention on certain events, originating with the thought of Descartes and culminating in that of Diderot, which stretched over the years (more or less) from 1650 to 1750, and which represent a particularly critical phase in the maturing, or modernizing, of the general concept in question.

In tracing the evolution of materialist science from its Cartesian sources to Diderot and his contemporaries, a definite method has been observed. This is to give, unlike many scholars who have already examined Descartes's influence on the Enlightenment, the fullest scope and weight to the testimony and other materials (in some cases unpublished) provided by actual eyewitnesses and participants, even when these latter are no longer remembered on their own merits.

PREFACE

In the historiography of ideas, the judgments of the "great," which in part trespass onto the future, must frequently be integrated with those of the "less than great," which are more firmly implanted in their own times, in order to avoid anachronistic distortion of the past. With all this, the present work does not pretend to an exhaustive treatment of its subject. It has been thought preferable, in a first discussion of so broad a topic, to range freely over the field, rather than to labor too intensively any one of its many areas. Moreover, several ramifications of the main problem, which have had to be considered somewhat summarily, in themselves suggest separate studies.

It is a pleasure to take this occasion to acknowledge the criticisms, suggestions, and encouragement that I have received from various persons. My chief debt in this regard is to Professor Norman L. Torrey, of Columbia University. I should like also to express here my thanks to other members of the Columbia Faculty: to Professors Nathan Edelman and Otis Fellows, of the Department of French; to Professor Charles Frankel, of the Department of Philosophy; and to Professor Marjorie Nicolson, of the Department of English and Comparative Literature. Finally, I would like to state my special gratitude to Professor John Herman Randall, Jr., for the very sympathetic and helpful interest that he has shown in my efforts.

A. V.

Cambridge, Massachusetts
October, 1952

CONTENTS

CHAPTER I

AN ASPECT OF THE CARTESIAN HERITAGE

Descartes, ce génie extraordinaire né pour égarer et pour conduire . . .

—Diderot, *Œuvres*, XIII, 371.

CHAPTER I

AN ASPECT OF THE CARTESIAN HERITAGE

THE point of departure of the present study is the truism that Descartes's philosophy, despite the spiritualist metaphysics on which it was claimed to rest, concealed the incipient germs of modern naturalism. Depending whether its explicit or implicit features are stressed, the Cartesian system can be made either the fountainhead of Malebranchean, Berkeleyan, Kantian, and nineteenth-century idealism, or the instigator of eighteenth-century *lumières*. Yet, the Enlightenment itself was an ideological event of great complexity and internecine conflicts. Only too often, in attempting to state the relationship of Descartes to the Enlightenment *in toto*, the former's doctrine has unavoidably been quintessentialized to an almost deceptive generality. Our intention, however, is to restrict the Cartesian influence on the philosophes to a special and definite subject: namely, the rise of materialism as reflected in Diderot's writings principally, and, in so far as a common outlook was manifested, in those of Buffon, La Mettrie, D'Holbach, Maupertuis. This materialist movement, moreover, will be considered with reference particularly to its origins in the history of scientific method, theory, and discovery. The role performed by Cartesian ideas must, therefore, be accordingly delimited. The specific problem to be examined is *how* Descartes's natural philosophy, and the general conception of science it entailed, made together for the genesis, in the next century, of a trend to scientific naturalism which culminated in the ideology of Diderot and certain of his contemporaries.

As preliminary orientation, one ought to review how the question of Descartes's ties with the Enlightenment has variously been treated by historians of literature and philosophy. Roughly, three broad conclusions have been reached: first, that a decisive break took place in doctrine and in spirit from Descartes to the philosophes; second, that a fundamental similarity is evident between the two bodies of thought; third, that the philosophes, dividing the Cartesian system as it suited their purpose, discarded the metaphysics and physics, but brought to full fruition the rationalist, critical method of Descartes.

Bouillier's *Histoire de la philosophie cartésienne* spoke of the activities of Voltaire, Diderot, Condillac, taken together, as the "philosophical revolution of the eighteenth century," tracing it mostly to the influence in France of the combined ideas of Bacon, Locke, and Newton, imported by the philosophes to stock their intellectual arsenal in the war on dogmatism.[1] But experimental science and empiricist psychology, thus transplanted by Voltaire and Condillac respectively, had apparently not long kept their original characters on French soil. From the 1740's on, materialist attitudes, as witnessed in differing ways by La Mettrie, Diderot, Maupertuis, Buffon, D'Holbach, began to appropriate an important part of the main stream of Enlightenment. In Bouillier's estimation, this occurrence was presumably the outcome of a Lockean-Newtonian complex of ideas with which certain philosophes, not suspecting the final result, had sought at first to unseat from power the Cartesian philosophy. Bouillier's opinion about the sources of French naturalism, somewhat modified, still enjoys wide acceptance. It persists also, to some extent, in the view of those who contend that the philosophes, by having repudiated Cartesian metaphysics and physics, cannot be regarded as the proper heirs of Descartes in any genuine sense.[2]

[1] Francisque Bouillier, *Histoire de la philosophie cartésienne*, 3e éd., Paris, 1868, II, 555-579.

[2] Cf., for example, Charles J. Beyer, "Du Cartésianisme à la philosophie des lumières," *Romanic Review*, 34 (Feb. 1943), pp. 18-39. The author,

Taine took a quite different approach. He maintained, contrary to Bouillier, that the natural and moral sciences of the eighteenth century were basically in accord with the *a priori* procedures of Descartes and his school. In vain had Condillac, Rousseau, Helvetius, Condorcet, and the *idéologues* claimed Bacon as their master. Even if their initial borrowings had been fresh to French minds, they had soon enough fallen into step with Cartesian tradition; after a minor reliance on experience, like the Cartesians they had abandoned it altogether.[3] To Taine it seemed that the philosophes' thinking derived from an essentially aprioristic conception of science, whose ideal was to "extract, circumscribe, isolate a few very simple and extremely general notions; then, leaving experience behind, to compare and combine these, and from the composite thus artificially obtained, to deduce by pure reason all the consequences that followed."[4] Such an evaluation deserves, in connection with our study, to be retained. Taine's thesis, as far as it goes, would suggest that the philosophes'

---

assuming like Bouillier that Descartes's system, necessarily an organic whole, was definitely rejected as such by the philosophes, searches for the indigenous roots of French *lumières* in the seventeenth-century currents of epicureanism, skepticism, and empiricism (Gassendi, Bayle, Fontenelle), which are said to have assimilated to their own ends the Cartesian rational method. This opinion, while valuable for its attention to the purely native sources of the Enlightenment, overlooks two points essential to the correct understanding of the latter's materialist phase. These are, first, the profoundly naturalistic implications of Descartes's science of nature, and secondly, its position of dominance, and correspondingly great influence, in France up to the 1730's.

3 Hippolyte Taine, *Les Origines de la France contemporaine: l'Ancien Régime*, 16e éd., Paris, 1891, p. 263.

4 *ibid.*, 262. A more recent analysis in some ways like Taine's may be found in Antonino Bruno, *Cartesio e l'Illuminismo*, Bari, 1949, esp. pp. 30-35. The author establishes the tie between Descartes and the philosophes in terms of their common anti-historicism, or their appeal to a set of absolute and eternal values which, as objects of pure reason, transcended experience of the historically contingent in the various spheres of human knowledge. This estimate, best applicable to the Enlightenment's deistic phase (Voltaire, Rousseau, D'Alembert), would not entirely fit, as will presently be seen, the materialist school's attempt, under Cartesian auspices, to synthesize the historical and the rational.

scientific naturalism, along with other phases of Enlightenment thought, was deeply rooted in Cartesian intellectual habit; that, notwithstanding the superposition of Lockean and Newtonian theories on it, the current of French science that gave birth in time to materialism remained docile, in a fundamental sense, to rationalist rather than experimentalist direction. This is, no doubt, oversimplifying the question; but there will be occasion later to dwell on the interactions and nuances between experimentalism and rationalism in the eighteenth century, and to note what compromises and syntheses were achieved, in theory and in practice, by the philosophes.

The third major solution of the problem should be mentioned, if only to differentiate it from the purpose of the present work. Brunetière effected a kind of adjustment between the two foregoing opinions. To do this, he had mainly to consult what the philosophes themselves had acknowledged about their debt to Descartes. Voltaire, Diderot, and D'Alembert, Brunetière held, were in reality the legitimate heirs of the Cartesian reform; but in order to develop the meaning latent in their ideological heritage, they had been obliged to disengage the philosopher's classic method, and its apotheosis of reason, from his specific scientific and metaphysical doctrines.[5] Descartes's criterion of rationality, temporarily checked by Jansenism and the preponderantly religious temper of the seventeenth century, had begun to triumph only around 1700, when it joined forces with libertine, free-thinking, and skeptic tendencies. Lanson brought to the support of Brunetière's thesis a truly impressive array of detail. What he called the "philosophical spirit of the eighteenth

[5] Fernand Brunetière, *Etudes critiques sur l'histoire de la littérature française*, 4e série, 1889; "Jansénistes et Cartésiens," pp. 111-178. An echo of Brunetière's thesis may be found in A. Dal Sasso, "La Influenza di Cartesio sulla formazione dell'Illuminismo francese," *Cartesio: terzo centenario del "Discorso del Metodo,"* Pubblicazione dell'Università cattolica del Sacro Cuore, Milan, 1937, pp. 227-238.

century" was traced to Descartes's twin standards of universal doubt and geometric logic—plus Montaigne's skepticism, Gassendi's epicureanism, Bayle, Fontenelle, Spinoza, Locke, Malebranche, and Newton, to mention only philosophical contributors.[6] So pervasive and intangible a thing as an "esprit philosophique" could hardly have been defined in a less comprehensive manner. Yet, in Lanson's researches the role of Descartes was inevitably obscured, if not volatilized, by overgeneralizing the rule of geometric reason abstracted from almost every other feature of his philosophy.

It is our belief that the broad problem posed by Bouillier, Taine, Brunetière, Lanson, and others, could best be treated by first particularizing the relations between Cartesianism and the Enlightenment; by assuming that, in their indebtedness to Descartes, irreducible differences might have existed among the philosophes. To this end, attention will here be fixed chiefly on one person—Diderot, despite the fact that the encyclopedist's mobile mind, like the weathervanes of his native Langres, shifted freely in the winds of thought. Nonetheless, in the veritable maelstrom of ideas that agitated his century, Diderot did represent, together with La Mettrie, Buffon, and D'Holbach, a definite ideology of scientific naturalism which, in the chapters following, will be distinguished from other types of materialism and other trends in science during the period. Moreover, in pursuing the line of descent from Descartes to Diderot within the confines of our study, it will not be necessary to keep apart, except in a few cases, the teaching of Descartes from that of Cartesianism, for these complemented each other in their interwoven influence on the philosophe.

The efforts of historians of philosophy to determine the sources of eighteenth-century French materialism have par-

[6] Gustave Lanson, "Origines et premières manifestations de l'esprit philosophique dans la littérature française de 1675 à 1748," *Revue des cours et conférences*, 1907-1909, *passim*.

alleled, to some extent, the contradiction of opinions already exhibited regarding Descartes's general contribution to the Enlightenment. Lange's *History of Materialism* was inclined to accept as very considerable Descartes's influence on La Mettrie, Diderot, and their associates. It was presumably the author of the *Principes* and *Traité de l'homme* who had impressed on materialist thinking "that stamp of *mechanism* which appeared most strikingly in De La Mettrie's 'L'Homme machine.' It was really due to Descartes that all the functions as well of intellectual as of physical life were finally regarded as the products of mechanical changes."[7] Despite this promising start, it was doubtless the lack of precise historical documentation that led Lange, elsewhere in his work, to regard the influence of Descartes as having been either exhausted around 1700 or "tamed" by the Jesuits shortly thereafter, and to seek in Newtonian tradition the sources of materialism and atheism among the philosophes.[8] Present-day scholarship has but imperfectly freed itself from the belief that it was mainly the introduction of Locke and Newton into France which, by some alchemy of ideas whose formula has, however, remained inadequate, there produced naturalistic science. Emile Bréhier, for example, although cautious about linking English sensationalism with the materialist attitudes of La Mettrie and Diderot, nonetheless leaves Descartes conspicuously out of the picture, and maintains, in another section of his book, that Newton and Locke ruled over eighteenth-century France like a kind of absolute duumvirate.[9] Bertrand Russell, in a recent work, recognizes momentarily that the evolution of certain Cartesian doctrines brought about during the Enlightenment a consistent materialism, but elsewhere designates, without serious reservations, as Locke's

[7] Friedrich Lange, *History of Materialism*, tr. E. C. Thomas, 2nd ed., London, 1879-1881, I, 243, and ff.

[8] *ibid.*, II, 15.

[9] *Histoire de la philosophie*, Paris, 1930, Tome II, fasc. 2, "Le Dix-huitième siècle," pp. 311-312.

heirs, "those of the French *philosophes* who did not belong to the school of Rousseau,"[10] which would of course include La Mettrie, Diderot, D'Holbach. Without questioning that Locke and Newton, in the respective spheres of empiricist psychology and experimental science, were in effect the masters of the age of *lumières*, our purpose will be to show, by establishing the necessary ties between Descartes and scientific naturalism, that English philosophy was at best secondary to Cartesian precept in prompting and molding a major segment of Enlightenment thought.

Patently, the philosophes did not always respect the intended meanings of ideas borrowed from the past. History is the realm of continuity, not of consistency; and this being so, Cartesian philosophy and Cartesianism are admittedly two different things. A materialistic rendering of the former would of course not be tenable in the light of present-day knowledge. Our problem being one primarily of ideological filiations, however, Descartes's philosophy will deliberately be viewed from the angle which affords the best understanding of the transformation—even deformation—it underwent to yield eighteenth-century naturalism. While it is beyond the limits of this study to fix the degree of faithfulness shown by the philosophes to whatever might be taken as a rigorous ex-

[10] *History of Western Philosophy*, London, 1946, p. 666. Ernst Cassirer, accepting the conventional linkage between Newtonianism and French materialism, has been led to judge the latter an inconsequential and accidental occurrence, owing no doubt to the difficulty of demonstrating its ideological alliance with the mechanistic experimentalism of Newton. Cf., *Die Philosophie der Aufklärung*, Tübingen, 1932, pp. 72-73: "Um die Mitte des Jahrhunderts ist diese Auffassung durch die Schüler und Apostel, die Newtons Lehre in Frankreich fand, durch Voltaire, durch Maupertuis, durch d'Alembert zur allgemeinen Geltung gelangt. Man pflegt die Wendung zum 'Mechanismus' und 'Materialismus' als ein charakteristisches Kennzeichen der Naturphilosophie des achtzehnten Jahrhunderts zu betrachten, und man glaubt nicht selten damit insbesondere die Grundrichtung des französischen Geistes erschöpfend bezeichnet zu haben. In Wahrheit handelt es sich jedoch in diesem Materialismus, wie er z. B. durch Holbach's 'Système de la Nature' und durch Lamettrie's 'L'Homme Machine' vertreten wird, um eine vereinzelte Erscheinung, der keineswegs typische Bedeutung zukommt."

position of the Cartesian system, the latter's materialist career had a definite logic of its own, deserving of attention. Above all, it is futile to dismiss the philosophes' interpretations as "a total incomprehension."[11] In the final estimate of any philosophy great enough to have had a history, abstraction cannot be made of what succeeding ages have been able to see in it. Diderot and his colleagues, by carrying to its ultimate conclusion one part of Descartes's doctrine—the mechanistic—and revising the remainder in accordance, were Cartesians in the proper sense of the word. The neglect up to date of this important phase of Descartes's influence is all the more surprising, when one considers that it represents a particularly effectual form in which Cartesianism has reached the twentieth century.

Metaphysical dualism was itself the first great step towards naturalism, paradoxical as this might seem.[12] By the distinction of two substances, *res cogitans* and *res extensa*, not only idealism was provided with a firm foundation: matter was raised to a revolutionary status by comparison with its former position at the rock-bottom of Aristotle's scheme of entelechy. In the history of philosophy, the more radical change is sometimes apt to be the lesser modification that continues to remain apparently orthodox (as did Descartes's metaphysics), rather than the overt innovation. The substantializing of matter on a footing equal to soul was just such a radical change; in the end, it could be expected perhaps to produce a deeper divergence from the past than, say, Gassendi's concurrent revival of Epicureanism. For, a substance is by definition something that gets on better alone than in pairs. While Descartes's system became, in consequence, an attempt to demonstrate formally the union of two substances by their concrete mutual functionings in man, the very mode

[11] Jacques Chevalier, *Descartes*, Paris, 1921, p. 343.

[12] Cf., Jacques Maritain, *Le songe de Descartes*, Paris, 1932, *passim.*

and denotation of such a union remained essentially unintelligible.

As a result, the period following witnessed three ingenious philosophies which sought, in diverse ways, to weld together the realms of matter and soul, or of natural science and theology, that had been left by Descartes in a precarious state of separation and virtual autonomy. These were Malebranche's doctrine of occasional causes; Leibniz's preestablished harmony; and, in a sense lost on the eighteenth century, Spinoza's pantheism, which was for long construed as merely a brand of atheistic materialism. Owing to this view of the *Ethica*, Spinoza's ideas were to be assimilated, in an understandably non-technical and adulterated version, to the naturalistic current, by Diderot among others. As for Malebranche and Leibniz, despite a restricted use made of certain of their scientific theories, it may be said that neither occasionalism nor the preestablished harmony had any appreciable impact on the philosophes. It was gravely in their disfavor that Malebranche and Leibniz, to prove the bond between physics and theology, had had to resort to unusual subtlety of argument. If the philosophes had thoroughly learned one lesson from Descartes, it was that truth was simple and recognizable by any well-directed mind—and they confessed an inability to make much sense of either Malebranche or Leibniz. Descartes's criterion of clear and distinct notions, with its metaphysical supports in the *Cogito*, served ultimately to discredit all hope of rendering coherent the Cartesian dualism. The philosophes were fond of saying that they had turned Descartes's own weapons against him. In the case of the materialists, this meant their using the rules of evidence to develop the physical at the expense of the theological portion of his system, the latter being what Malebranche and Leibniz had endeavored above all to save. The Cartesian *res extensa*, and the new standing it accorded to Nature as an entity, were allowed to unfold their full potentialities during the Enlightenment. Against the concept of matter as substance, Diderot's group was not to invoke disparagingly the test of clear and distinct ideas.

A curious text, probably from the pen of Grimm, epitomizes the character of the continuity that led from the *Cogito* to the naturalistic position. The procedure practiced in the *Discours de la méthode* of stripping experience down to its inexpugnable core had been transposed, by the eighteenth century, to a less philosophical, but more concrete, context: reduction to the most immediate became, for the philosophes, reduction to the most primitive. Natural man, assumed (rather too lightly) to be free of the prior opinions, prejudices and conventional notions of civilized society, came to figure as a kind of incarnation of the hero in the *Discours* who had tried, all too hypothetically, to divest himself of all tradition, education, and sensory determinants. But the reflective savage, it seems, would have employed the method of the *Cogito* to better advantage than Descartes who, after all, had been subject to theological control:

"It is evident that a man born with a genius for meditation and raised among a well-tempered, savage people, or cast on a deserted isle far from our opinions, reveries, and metaphysical as well as theological absurdities, would begin his philosophy by means of the same principle used by Descartes. But he would not at all arrive at any of the results of the Cartesian system. *His* philosophy would be clear and precise. He would say: I think, therefore I exist. But he would not say: There is within me a being that thinks. . . ."[13]

And the passage proceeds to extract from the primary certainty of the *Cogito* a plainly materialist conception of man and the world. The reasoning involved is not put in perfect

[13] Grimm, Diderot, Raynal, Meister, *et al.*, *Correspondance littéraire, philosophique et critique*, éd. M. Tourneux, Paris, 1878, VI, 361. Unless otherwise stated, translations in this work are mine.

logical form: its interest for us lies, rather, in the abridged version it offers, retrospectively, of the historical process which, going continuously from Descartes's time to that of the philosophes, had by stages converted dualistic metaphysics into scientific naturalism. Starting from the *Cogito*, where the surest experience and clearest idea coincided in the disclosure of reality, the series of propositions leading to materialism are as follows: (1) however certain it is *that* one thinks, Descartes's method requires an avowal of ignorance concerning *how* one thinks; (2) thought, considered from this aspect, is no more clearly conceivable than movement; (3) the succession of images in the mind indicates the presence of an outside world; (4) the external world cannot be conceived to have a beginning; (5) matter itself cannot positively be defined as incapable of thought; (6) that matter can think is more easily conceivable, and incurs less general difficulties, than the dualist postulate; (7) therefore an infinite material substance thinks, and dualism is inadmissible.[14]

Desultory as this "chain of reasons" might appear, it contains a valuable hint of the permutation worked in Cartesian metaphysics by the naturalistic school. The *Cogito* had dealt essentially with the validity of thought, not its spirituality; the proof that the thinking self exists did not imply necessarily the status of a *res cogitans* as substance. To eighteenth-century materialists, for whom Kant had not as yet translated the meaning of the *Discours de la méthode* into an idealism of the subject, Descartes's realist intuition appeared to have for its object mind *quâ* matter. Diderot was well aware, on this point, of Hobbes' paraphrase of the crucial sentence that underlay Cartesian metaphysics: "I think, therefore matter can think."[15] Furthermore, the existence of God, directly deducible from the *Cogito*, had primarily served

14 *ibid.*, 361-362.

15 Denis Diderot, *Œuvres complètes*, éd. Assézat et Tourneux, Paris, 1875-1877, xv, 98-99, article "Hobbisme."

Descartes to authenticate human reason in its quest for scientific certainty of the external world. The Cartesian God who guaranteed the veracity of clear and distinct knowledge about the realm of matter, far from presenting an obstacle to naturalistic science, was the one God in whom actually Diderot and his circle, by their complete faith in the mechanistic formula, may be said to have believed, and like true believers, tacitly. When Diderot objected that God would be a deceiver if man's reason were unreliable,[16] he was in effect expressing the analogue of Descartes's exorcism of the *malin génie* by the divine attributes.

The key-change made by materialist Cartesianism in the philosopher's metaphysics was the negation of the dualist premise, and with it the possibility of the soul's immortality. In itself, the distinction of two substances had not been, strictly speaking, a deduction either from the *Cogito* or from the attributes of God; as Descartes made explicit, it had been affirmed simply as a clear and distinct notion.[17] But its clarity and distinctness soon became suspect owing to the obscurity and confusion it incurred regarding the union in man of body and mind. Aside from the attempts (not wholly convincing) of Leibniz and Malebranche to resolve the dilemma within the dualist framework, several alternatives remained: to unite mind and matter as modes of God's being (Spinoza); to ignore the metaphysical problem (Locke); to make all things mind (Berkeley); or to make all things matter. Of these, the last-named alone was able to claim that its denial of the real distinction between the corporeal and the spiritual was imposed specifically by the requirements of Descartes's natural science itself. If, as numerous scholars have maintained,[18] the

16 *ibid.*, I, 159.

17 Letter dated March 1638: "De cela seul qu'on conçoit clairement & distinctement les deux natures de l'ame & du corps comme diverses, on connoist que veritablement elles sont diverses." *Œuvres de Descartes*, éd. Adam et Tannery, Paris, 1897-1913, II, 38.

18 Among others, Liard, Adam, Laberthonnière, Gilson, Maritain.

success of his physics was truly Descartes's principal concern, the materialist revision of his metaphysics in the interest of the former remained, in a special sense, faithful to the spirit, if not the letter, of Cartesian doctrine.

Regarding the realm of matter, Descartes had left entirely open the way to progressively deeper inquiry, holding out the promise of human progress. Whereas his spiritualist metaphysics remained relatively static, his natural philosophy was, so to speak, dynamic. What eventually happened to Descartes's ideas in the course of history was largely the outcome of this disparate situation. He had failed to say how the apologetic features of his system were to profit its readers; even in psychology, the path to greater knowledge and to moral betterment lay through the physiological investigations of the *Traité des passions*, rather than through the ontological speculations of the *Méditations métaphysiques*. The program of exploring nature was justified, basically, by the multitude of benefits it would secure to man on the practical level. The theme of conquering the physical world, which had spurred the efforts of the Renaissance, was given a twofold utilitarian valuation by Descartes. First, scientific activity found immediate, personal application as both the content and purpose of what Gouhier has called "the life of the intellect."[19] Second, the closing pages of the *Discours de la méthode*, in stressing the incalculable uses of the new physics for medicine, mechanics, and morality, broke with the verbalistic science of the Schoolmen, among other ways, by making preëminent the scruple of public utility: "Instead of that speculative philosophy that is taught in the schools, one might find a practical one," etc.[20]

The Enlightenment strove to carry out in its full sense the messianic overtone of Descartes's natural philosophy.

[19] Henri Gouhier, *Essais sur Descartes*, Paris, 1937, p. 6. See also, pp. 63-65, 114-117, 122-123.

[20] Descartes, *Discours de la méthode*, texte et commentaire par E. Gilson, Paris, 1925, "Part vi," pp. 61-63.

The *Encyclopédie*, in its article on "Cartésianisme" by the abbé Pestré, recognized that the philosopher had made the crowning goal of his scientific program the elimination of "an infinity of maladies, and perhaps even of the infirmities of old age."[21] In the *Eloge de Descartes* that won the prize of the French Academy in 1765, and earned the applause of Voltaire and Grimm's *Correspondance littéraire*, Thomas summarized the Cartesian ideal: "Let all scientists unite their forces. Let one group begin where another has left off. Thus, by joining together the lives of various persons and the labors of several centuries, a vast depot of knowledge will be formed, and nature will at last be subjected to man."[22]

In phrasing the utilitarian credo which, for Thomas, was to be "la grande idée de Descartes," the language of the *Discours de la méthode* had coincided with that of another, earlier rebel against Scholastic futilities—Francis Bacon. Now, the fact is that when Diderot, in the *Pensées sur l'interprétation de la nature* of 1754, proclaimed the same hope of making nature the servant of humanity, he chose to leave the reader with the impression that Bacon, not Descartes, had invoked this attitude. But the scientific optimism which Diderot had acquired merely by being born at the proper time and place owed its main impetus, indisputably, to the powerful domination by Cartesianism of French thought up until the 1730's. Diderot had read and reread the *Discours de la méthode*, with its eloquent proposal to render mankind "comme maîtres et possesseurs de la nature," long before familiarizing himself with the works of Bacon. Why was it felt necessary, then, to bring Bacon so prominently into the picture, and to exclude Descartes? Here we are faced with a problem typical of an entire class. It will be discussed presently why the philosophes were habitually unwilling to ac-

[21] Diderot, D'Alembert, *et al.*, *Encyclopédie, ou Dictionnaire raisonné des sciences, des arts et des métiers*, art. "Cartésianisme," II (1752), 719.

[22] Antoine-Léonard Thomas, *Eloge de Descartes*, Paris, 1765, p. 110n.

credit to Descartes much that they had in all probability learned from him. At the moment, however, we need not be deceived. When Diderot voiced the aspiration of controlling nature for practical human ends, and advised that progress made by scientists be presented to the public with special reference to its utility,[23] he was seconding Descartes and counseling the precise strategy employed in the *Discours*. For both of these, it is true, he had later found confirmation in Bacon.

That physics should be made to procure benefits to man through medicine and the mechanical arts was something in which any Schoolman or Jesuit, among the contemporaries of both Descartes and Diderot, would have concurred. But Descartes had said more. The three branches that sprouted from the Cartesian tree of knowledge included, not only medicine and technology, but also morality. In attaching the sphere of morals to the practice of a method for the advancement of science, a radical notion was coined and given currency; its tremendous possibilities have not today been exhausted. It is a curiously striking fact that the *Discours*, while ostensibly expounding a method suited to natural science, discussed and evaluated this persistently in terms of its importance for the conduct of life. Descartes seemed to be offering simultaneously a method for the explanation of physical events and for the exploration of man's nature and earthly destiny. An earlier title for the *Discours* had emphasized just this new and promising theme: *Projet d'une science universelle qui puisse élever notre nature à son plus haut point de perfection*, etc. The discrimination of the true from the false that the rules of evidence guaranteed, was linked directly to Descartes's wish "to see clearly in my actions, and to walk with assurance in this life."[24] As for theology, which was paid profound but somewhat nominal respect, its

23 Diderot, II, 9, 19.
24 Descartes, *Discours*, "Part I," p. 10.

role was presumably for the conduct of the after-life; superhuman powers or the intervention of extraordinary grace was needed, thought Descartes, to make headway in questions of faith. But with sound physical knowledge substituting for the musty accumulations of tradition, Descartes believed it possible to live in the best manner in the present world: "I held firmly that by this means I would succeed in conducting my life much better than if I had built only upon old foundations."[25] More exactly, the connection of the new science with ethical meliorism—or the specific tie between physics and happiness—was to be established by the function of medicine (understood in its broadest sense): "for even the mind depends so much on the temperament and disposition of the bodily organs that, if it is possible to find some means by which men might commonly be made wiser and abler than they have ever yet been, I believe it is in medicine that it ought to be looked for."[26]

As is well known, Descartes delivered this revolutionary piece with extreme caution and various reservations in order to escape censure. But despite the "provisional morality" and the tactful handling of religious sensibilities in the *Discours*, the author's message proved to be a compelling invitation to a future age to effect deep changes both in moral doctrine and in the attitude towards the passions.[27] Inspired by Descartes, the philosophes abrogated much of the authority, in the study of man, assumed by the Church at the expense of natural science. La Mettrie's *l'Homme machine* acknowledged, in this respect, Descartes's achievement in having furnished one of the bases on which eighteenth-century

[25] *ibid.*, "Part II," p. 14. [26] *ibid.*, "Part VI," p. 62.

[27] Voltaire recognized Descartes as a precursor of the rehabilitation of the passions carried on during the Enlightenment. The *Supplément aux œuvres en prose* remarked that the philosopher had written on the passions, "mais pour les approuver, et qu'il les trouve toutes bonnes, surtout l'amour." *Œuvres complètes*, éd. Moland, Paris, 1877-1883, XXXII, 596. It was, similarly, the sexual passion that Diderot and La Mettrie seemed most interested in vindicating against Christian asceticism.

materialism was to build boldly and confidently: "There are as many different minds, characters and mores as there are temperaments. Even Galen knew this truth, which Descartes . . . has made so much of, to the point of saying that Medicine alone is able to change thinking and behavior together with the body."[28] As the moral guide and benefactor of society, the investigator of nature took over a large part of the duties and prerogatives—admittedly with some inconvenience—that had belonged perennially to religion. The deeds of saints, past and present, were relegated to the ignorance and misguided endeavor of an "un-enlightened" era. "It is, properly speaking, to be worth nothing," Descartes had said, "if one is not useful to somebody."[29] And it was the cultivation of natural science, according to the *Discours*, by which not only one's contemporaries, but more important still, posterity could best expect to be served and improved. Eulogizing on Descartes's accomplishment, Thomas stated that not merely had the great philosopher asserted the fecund idea of the unity of science, but had related all the parts of his universal system to man.[30] The potent themes animating the movement of Enlightenment, such as human progress, moral perfectibility, the obligation to posterity, thus coincided with, and forcefully expedited the materialist orientation in science which La Mettrie declared he had found in Descartes.

The contemplative rationalism of the Scholastics, tending to place the *summum bonum* in the timeless (and often untimely) speculations of the metaphysician, was converted by Descartes into a practical rationalism whose chief aim was the temporal betterment of man's estate and capabilities, and whose implement was physical reality instead of the syllogism. After Descartes, at least two vast attempts were made to

[28] Julien Offray de La Mettrie, *Œuvres philosophiques*, Amsterdam, 1774, III, 7.

[29] Descartes, *Discours*, "Part VI," p. 66.

[30] Thomas, *Eloge de Descartes*, p. 63.

solve again the general problem of morality on the metaphysical plane: these were the pantheistic determinism of Spinoza, and the Leibnizian optimism based on the principle of sufficient reason. It is an indication, however, of how well Descartes's lesson had penetrated that the philosophes treated with impatience, or regarded as pernicious, the ethical systems of Spinoza and Leibniz. In relying as moral philosophers on purely rationalist schematizations having little direct utility in the world, Spinoza and Leibniz seemed to be "Scholastics" bent on undoing what Descartes had hoped the future would carry to its expected conclusion. By contrast, Montesquieu's *Lettres persanes* gave the philosophe's standpoint, which was a logical enough extension of the teaching in the *Traité des passions*: "Nothing is so afflicting as those consolations based on the necessity of evil, the uselessness of remedies, the inexorability of fate, the order of Providence, and the misfortune of man's condition. . . . It would be much wiser to free the mind from such reflexions, and to treat man as an emotional rather than a rational being."[31]

This came to mean, for the Enlightenment, the need for treating man as an adjunct of what was called, with a variety of meanings, Nature. The term itself was convenient particularly by its vagueness. Nature was the vade mecum in which the philosophes found, diversely, their standards of the true or the good; it was the amorphous concept by means of which the Newtonian deist might commune with God through the spectacle of the heavens, or with which Rousseau and his adepts might hope to reconcile sentiment with logic. Here, however, the word must be given a more restricted and "Cartesian" connotation. Nature will be considered as the substantive object of natural science. For materialists such as La Mettrie, Diderot, and D'Holbach, Nature, whatever else it might have been besides, was essentially the *ensemble* of

[31] Charles de Secondat, Baron de Montesquieu, *Lettres persanes*, XXXIII, in *Œuvres complètes de Montesquieu*, éd. Laboulaye, Paris, 1875-1879, I, 134.

particles—their shapes, sizes, and arrangements—that composed the physical universe, together with the general laws governing their complex behavior. The exponents of scientific naturalism in the eighteenth century, in so far as they remained consistent with their own doctrine, intended ultimately to be able to refer all problems of a psychological, moral, or social character to Nature so defined, the Nature of the physicist and biologist.

In comparing Descartes and the philosophes, actually two quite unequal things will be juxtaposed. A system of philosophy in the proper sense will be set against a rather loose complex of ideas lacking internal rigor. The naturalism of La Mettrie, Diderot, D'Holbach might best be designated as an ideology, that is, an unsystematic grouping of interrelated notions and attitudes, suitable for wide dissemination and the enlisting of public opinion in the pursuit of practical ends. For the philosophes by and large, natural science, in addition to being the means of ascertaining objectively truths of a certain category, was an effective instrument of ideology. But great as the historical transition from a philosophy to an ideology must understandably be, here too it was Descartes who had showed the way that the others followed.

The new method outlined in the *Discours* was offered as something that anyone, temperamentally fit and able to grasp a geometrical demonstration, could acquire for his own use; its main attraction was that no special erudition or rare intellectual gift was needed for its mastering. True, Descartes had given some precautions against the use of his method by all and sundry, but this was not to be the cue on which posterity acted. *Bon sens*, or the faculty of rational judgment, being the most equitably shared thing on earth, it became possible for all who adopted the few simple rules of the *Discours* to "distinguish the true from the false" for themselves. This did not mean, as Descartes was careful to insist, that unusual qualities of mind and imagination were not required, besides the right method, for making important discoveries and advances in the sciences. It did mean, however, that anyone could independently understand them and estimate their soundness by applying in turn the same method. If Descartes published his Meteors, Dioptrics, and Geometry in French, he obviously held that, by the aid of the prefatory essay on Method, his scientific opinions would be for the average intelligent reader "so simple and so in conformity with common sense as to appear less extraordinary and foreign than any others that might be entertained on the same subjects."[32] The original title of the *Discours* perhaps clarifies still better Descartes's intention to explain, in its pages, the elements of a "universal science . . . in such a manner that even those who have not studied may understand them"; a general hope which the preface of the *Principes de la philosophie* reaffirmed.[33] Because the new method that was to reward the expert investigator with unprecedented headway in natural science was the same method which, also, would render such science accessible to others less expert, the *Discours* succeeded in creating a "community of interest," unknown earlier, between the scientist and the educated public. Science could be expected, in consequence, to be the business of all cultivated and thinking persons; its fundamental appeal became one of "popularity," as opposed, for example, to history, classical languages, law, etc., which called for long years of preparation and particular talents. In natural philosophy, claimed Descartes, one had only to start from the clearest and simplest notions possessed in common by all minds, in order to proceed step by step through a chain of self-evident deductions to the remotest, most complex conclusions.

[32] Descartes, *Discours*, "Part vi," p. 77.

[33] Descartes, ix, 12: "I'ai pris garde, en examinant le naturel de plusieurs esprits, qu'il n'y en a presque point de si grossiers ny de si tardifs qu'ils ne fussent capables d'entrer dans les bons sentimens et mesmes d'acquerir toutes les plus hautes sciences, s'ils estoient conduits comme il faut."

Descartes's definition of the character of scientific knowledge thus anticipated, if it did not actually make inevitable, the philosophes' use of science as a most valuable tool of ideology. The eighteenth century teemed with all sorts of manuals on physics, astronomy, biology, zoology, botany, and the like, whose chief and earnest purpose was to put their contents "à la portée de tout le monde." The authors of these popularizations (which often took the form of natural theologies), such as the abbé Pluche, Père Regnault, Charles Bonnet, Père Paulian, and others, seemed unaware that the spread of scientific information and mental habit might assist in the progress of naturalism, and give rise in time to a cult rivaling the established, revealed one. Be this as it may, the philosophes turned to good account their contemporaries' taste for scientific vulgarization, and were themselves genuinely bound to the Cartesian ideal of making science available to all who would learn the correct method. The *Encyclopédie* and Buffon's *Histoire naturelle* were monuments to this ambition. When the philosophes had come up against the difficulty of making Newtonian physics, which did not follow the deductive pattern, accessible to the rank and file of their readers, Voltaire had confessed, not without regret, that "very few persons read Newton, because one must be very learned to understand him."[34] Voltaire did what he could to remedy the situation with his *Eléments de la philosophie de Newton.* Diderot, too, complained about what he regarded as "affected obscurity" in the *Principia Mathematica.*[35] But the philosophes persevered in the hope of eventually harmonizing the content of Newton's, or anyone else's, system with the methodological form of Cartesian science. Diderot went on to say in words that recall the *Discours de la méthode*: "Let us hasten to render philosophy popular.

[34] Voltaire, *Lettres philosophiques,* éd. Naves, Garnier, Paris, "Lettre XIV," p. 74.

[35] Diderot, II, 38.

If we wish philosophers and scientists to make advances, let us approach the people from the point which the philosophers have reached. Will some say that there are certain works which can never be placed within reach of ordinary minds? If they say so, they only show themselves ignorant of what a good method and long habit may accomplish." D'Alembert's expression of the same general belief in the "Discours préliminaire" of the *Encyclopédie* disclosed even more explicitly its Cartesian antecedents: "It is perhaps true to say that there is hardly a science or art in which, if necessary, the most limited intelligence could not be instructed by the use of good logic; for there are very few sciences and arts whose laws or propositions cannot be reduced to simple notions, and arranged in so immediately perceptible an order that the chain would not be broken at any point."[36] Descartes's conception of science as a deductive procedure within the comprehension of average, properly-trained minds was destined to play, during the Enlightenment, a decisive role in the formation of scientific naturalism as an ideology.

There is admittedly a great difference between a philosopher and a philosophe. But hardly any philosophical system had tended so much as Descartes's to attenuate this difference. Owing to the inherent characteristics already described, Cartesian thought had shown an inclination, during its founder's last years, to become "worldly," thereby initiating a secularization of philosophy and science that reached its apogee in the following century and has perhaps not been duplicated since. From 1644 on, Descartes had recruited followers primarily from the class of *honnêtes gens*, entrusting the fate of his ideas to such persons as Chanut, Clerselier, the abbé Picot, the Palatinate Princess Elizabeth, and the Earl of Newcastle; death had unseasonably interrupted his conversion of Queen Christina of Sweden. During this period, both the *Principes de la philosophie* and the *Méditations*

[36] Jean Le Rond d'Alembert, *Œuvres complètes*, Paris, 1821-1822, I, 34.

métaphysiques appeared in French translation. But what Descartes did not live to continue, his principal disciples took up and carried out. A most distinctive feature of Cartesianism in the latter part of the century was its marked success on the level of polite society. The correspondence of Madame de Sévigné, like conversations at the time in the salons of the marquise de Sablé and the duchesse du Maine, attested the vogue of Descartes's ideas in such circles.[37] It is known, moreover, that two leading proponents of Cartesianism, Rohault and, after him, Régis, gave public lectures on the new philosophy in order to gain for it a wider acceptance. Their audiences—where one had to come early to be sure of a seat—consisted of members of both sexes and of all professions. Already Descartes's influence was working steadily towards shaping the philosophe.

In the Enlightenment, the art of philosophizing formed an intimate part of social practice at its best. This meant not merely that bourgeois and nobles should be interested in philosophy, but that the latter should descend to the high-water mark of public intelligence. Descartes's method had brought about the social education of the philosopher, which was necessary before the philosopher could educate society. Descartes had personally exemplified something fairly novel for the seventeenth century: the gentleman-philosopher. By a simple amalgamation of terms, time was to produce the philosophe, a type to which anyone might honorably aspire, from commoner to monarch. The school of La Mettrie, Diderot, and D'Holbach exploited this situation deliberately and profitably. About the codification of their ideology, Voltaire observed: "Who is the dangerous author? It is he who is read by the idle members of the Court and by the ladies. . . . The author of the *Système de la nature* has enjoyed the advantage of having himself read by scientists, by the ignorant,

[37] Cf., Bouillier, *Histoire de la philosophie cartésienne*, I, 437-444.

by ladies. . . ."[38] Eighteenth-century materialism, while not a philosophy in the technical sense, was perhaps not too inconsistent with the example of Descartes.

Indicative of the change in the meaning of philosophy from one century to the next was the mode of expression that serious thought took among the philosophes. Whereas expositions of truth had once customarily been confined to bulky dissertations, treatises, summae, and the like, the age of *lumières* stated its deepest convictions in the context of a dialogue, tale, or some such entertaining genre. This was a necessary act for creating and maintaining an ideological atmosphere. But here again Descartes would seem to have foreshadowed the Enlightenment. Late in his career, he had shown a frank interest in the advantages of the dialogue-form. Baillet's *Vie de Monsieur Des-Cartes* furnishes most pertinent information about the philosopher's intentions in this regard: "Descartes appeared to favor the art of the dialogue, especially in the last years of his life, as a means of presenting his philosophy more agreeably. . . . He had begun his *Recherche de la vérité* in this dialogue-form. . . . He had even proposed to treat in the same manner his *Méditations* and *Principes* since his second trip to France; and M. Clerselier had promised Père Poisson that he would carry out this task. . . ."[39]

Whether these remarks mean that Descartes planned to incorporate into the unfinished *Recherche de la vérité* the subject-matter of the *Méditations* and *Principes*, or wished merely to set to dialogue his metaphysical and physical doctrines, the project was of course never executed. Yet, one might suggest that Fontenelle's *Entretiens sur la pluralité des mondes*, which "expounded" vortex physics in a series of elegant nocturnal conversations with an aptly-chosen *mar-*

[38] Voltaire, *Dictionnaire philosophique*, éd. Naves et Benda, Garnier, Paris, I, 394.

[39] Adrien Baillet, *La Vie de Monsieur Des-Cartes*, Paris, 1691, II, 475.

quise, both typified the milieu into which Cartesianism had spread and was a partial fulfillment, whether its author realized it or not, of Descartes's proposal to put his system into dialogue. Fontenelle has been described as a foremost figure in the popularization of science, as the wittiest propagator of Cartesian physics, and as the forger of the link between science and ideology that the philosophes were later to put to wider use. Actually, in the trend to secularize both science and philosophy that covered the years, roughly, from 1650 to 1750, Fontenelle was an intermediary between what Descartes himself had begun and what the philosophes were to complete.

The dialogue-form, in which Cartesian thought tended to culminate, represented a sociable and worldly conception of philosophy. Descartes would appear to have adopted, for the probing and presentation of the problems that occupied the whole of his life, a technique which went back, by way of the Italian Renaissance, to Plato. The vogue of the philosophical dialogue in eighteenth-century France—whose manifestations ranged from the *Lettres persanes* and the colloquies of Voltaire to manuals on the mechanical arts, philosophy, and the natural sciences—can thus be said to have started from one of the terminal points of Descartes's great reform. But against this background, Diderot stands out as the example of an unusually effective and purposeful *dialoguizing* of ideas. Rather than employ the *entretien* as an artificial mold indifferent to its ideational content, Diderot consciously exploited the role of dialectic, or the contrapuntal exchange of opinions, as a method for exploring, developing, and communicating to others the principles and implications of scientific naturalism. The tendency is present as early as his *Promenade du sceptique* of 1747. In 1749, the crucial episode of the *Lettre sur les aveugles* involving Saunderson and his interlocutors—which contains already the germ of Diderot's naturalism and transformism—causes the mode of expression to fall abruptly into dialogue. The *Interprétation de la nature*, by its rapid changes of subject and frequent digressions, makes the impression, rather, of an interrupted monologue. But in the *Rêve de D'Alembert*, the substance of Diderot's materialist system is set forth through the dialectical development (without for that reason being, as some have construed it, a "dialectical materialism") of which the philosophical conversation, when authentic, is the most appropriate vehicle. The result was that the ideology in question, at its best in Diderot, remained not only sociable in character and tone, but acquired from the technique of the colloquy a certain mobility and expansiveness inseparable from its meaning. In this regard, the import of scientific naturalism among the philosophes was to consist, not so much in the elaboration and defense of a dogma or doctrine, as in the removal of obstacles to a progressively closer scrutiny of the physical universe of which, generally speaking, man was held to be an intimate, uniform part.

Quite aside from its being in dialogue-form, Descartes's *Recherche de la vérité* suggests another and more important relationship to eighteenth-century practice, in particular to the full-scale offensive against authoritarianism mounted by the materialist school. Although the intellectual interests and general tenor of the *Recherche* seem specially consonant with Descartes's preoccupations after 1644, the dating of the work is still a controversial issue.[40] For the purpose at hand,

[40] The existing data, limited to Baillet's testimony and to inference from the use made of Descartes's manuscripts in Clerselier's possession, would indicate a late date for the *Recherche*. G. Cantecor, however, in "A quelle date Descartes a-t-il écrit 'La Recherche de la Vérité,'" *Revue d'Histoire de la Philosophie*, II (1928), pp. 254-289, argued, on the basis of a subjective appreciation of its style, that the work should be assigned to Descartes's youth (ca. 1628). Answering this, Henri Gouhier has made clear how closely the subject-matter of the *Recherche*, in ideas as in tone, fits into the picture of Descartes's interests and occupations around 1648. *Revue d'Histoire de la Philosophie*, III (1929), pp. 296-320. Gouhier's analysis is developed in greater detail by Ernst Cassirer: *Descartes*, Stockholm, 1939, pp. 118-176.

however, the problem of chronology may be ignored, since it is the contents of the dialogue that are of primary concern. The significance of the *Recherche de la vérité*, in its extant form a fragment of what was to be a work consisting of two books,[41] has perhaps not yet been fully realized. Not only is it a valuable key for interpreting the integral sense of Cartesian philosophy,[42] but it is an indispensable document for understanding the evolution of certain attitudes and ideas from Descartes to the philosophes. The *Recherche de la vérité* illustrates perfectly those features of Cartesianism, already referred to, which led to the secularization of science and philosophy, and to the creation of the type of ideology best exemplified by the eighteenth-century materialists. If the *Recherche* is a late composition, it apparently was meant to sum up the mundane orientation of Descartes's intellectual activity in the years shortly before his death. If it was written earlier, it may be taken as evidence of an inclination inherent to Cartesian thought which, however, did not assume predominance in the philosopher's mind until around 1648.

The full title of the fragmentary dialogue, seeming to predict a stage of Cartesian influence still far in the future, transports us suddenly onto the eighteenth-century scene: *Recherche de la vérité par la lumière naturelle, qui toute pure, & sans emprunter le secours de la Religion ni de la Philosophie, determine les opinions que doit avoir un honeste homme, touchant toutes les choses qui peuvent occuper sa pensée, & penetrer jusque dans les secrets des plus curieuses sciences.* The scope of the work, arrestingly bold for the time, was apparently without limit or limitation. Descartes states at the outset that for the communication of his philosophy to the general reader he has found no style so fitting "as that

41 Descartes, x, 491.

42 Cassirer, *op. cit.*, p. 119, writes of the *Recherche de la vérité*: ". . . es gibt keine zweite Schrift—selbst den 'Discours de la Méthode' nicht ausgenommen,—die so sehr wie diese geeignet wäre, uns einen Einblick und eine erste Einführung in die Gedankenwelt Descartes' zu verschaffen."

of these polite conversations, in which each person discloses in a familiar way to his friends the very best portions of his thought."[43] The hero of the *Recherche* is a certain Eudoxe, symbolizing *bon sens.* Of mediocre intelligence but eager to know, he expounds for the benefit of his hearers, two "rares esprits"—one by erudition, the other by subtlety, the gist of his metaphysical and physical opinions, which happen to coincide with those of Cartesian philosophy. A precursory abstract of the dialogue's unfinished contents reveals that they were to include Descartes's method, metaphysics, natural science, and ethics—in short, the whole of his system. The author's meaning is manifest: to form sound views in philosophy and physics, an average mind is self-sufficient, provided only that the right method is scrupulously applied. Descartes's "lumière naturelle toute pure" would seem already to be diffusing itself into the "lumières" of the freethinker. Whereas the *Discours* remained somewhat esoteric because it discussed an intensely personal adventure of the intellect, the *Recherche* emphasized the didactic possibilities of the new methodological discovery, and foreshadowed thereby what a later age was in fact to learn from Cartesianism.

But the title of the dialogue appears to say a good deal more. The *honeste homme,* in deciding for himself all questions of importance in this life, is expected specifically to shun, not only the philosophical technicalities of the School, but the counsels of religion as well. Such a revolutionary program for the education of the leisured classes defines Descartes's contribution to the burgeoning of that "esprit philosophique" against which, a century or so after, the abbé Nonnotte, seeing professional and amateur philosophes everywhere, was to deliver his jeremiad:

"People of that type are to be found nowadays in all gatherings, all walks of life, and all forms of society. One

[43] Descartes, x, 498.

finds them in the field of law, in the Academies, in official posts of all sorts, in schools teaching all kinds of arts and sciences. Everybody wants to be a philosopher; to reason, judge, decide about everything in a philosophical manner."[44]

If Descartes had dared in the seventeenth century to hint at such an eventuality, it was most likely because the opinions actually expressed by Eudoxe, establishing God's existence and the soul's spirituality, happened to conform satisfactorily with the position of the Church. However, the dialectical method of inquiry illustrated by the *Recherche* (whose delayed publication in 1701 came none too late), was one that could just as well yield unorthodox results. In fact, the title itself suggests the possibility of quite different things in the offing. Descartes states, in the way peculiar to the *Discours de la méthode*, that the same rationalist procedure is valid for scientific investigation and for the resolution of all problems outside of science. The "secrets of the most curious sciences" compose a consistent, unified body of knowledge with "all that could occupy the thought" of an educated human being. By the exclusion of both technical philosophy and theological precept from the organization of such a practical composite of knowledge, Cartesian doctrine would itself seem to have glimpsed, by an internal motivation, the same implications that certain philosophes were to recognize in it and to develop.

Almost a century passed before the radical approach of the *Recherche de la vérité* became the point of departure for the scientific naturalism of Diderot's generation. The marquis d'Argens remarked cleverly that "fifteen years after the printing of Descartes's works, ladies reasoned much more sensibly in metaphysics than three-fourths of the nation's theologians."[45] The moral of this was that theology, in the

[44] Abbé Claude-François Nonnotte, *Dictionnaire philosophique de la religion*, Besançon, 1774, I, xxxi-xxxii.

[45] Jean Baptiste de Boyer, marquis d'Argens, *Lettres juives, ou*

narrowly religious sense of the word, was thereby rendered superfluous to the conduct of life. It followed from Cartesian premises that each person, when properly instructed in the usage of reason, should in time become, to some extent, his own scientist, philosopher, and theologian. This was the ambition to which the literary output of the "côterie holbachique" appealed, and which it promised to satisfy. Descartes had more than intimated that the penetration of the secrets of physical nature would furnish a corresponding competence in all other problems of moment to man. The same theme, considerably magnified, in the teachings of Diderot, La Mettrie, Buffon, D'Holbach was seized upon and turned to ridicule by their critics. Although the wisest of the materialists, Diderot himself, saw the infructuousness of dogmatically referring all moral and social questions to the laws of matter in motion (as attested by his satire, in *Jacques le fataliste*, on the physical-determinist conception of human values and destiny), the naturalistic school impressed its contemporaries with the revolutionary role it gave to the sciences of nature in the task of studying, understanding, and guiding humanity. What scientific naturalism meant to the eighteenth century is illustrated, for example, by the description of it given in a typical (and perhaps the most successful) attack on Diderot's group: Moreau's *Nouveau mémoire pour servir à l'histoire des cacouacs*. The caricatural exaggeration of this diatribe reveals the salient features of the materialist ideology. The land of the "cacouacs," that is, the camp of scientific naturalism, takes as its gospel the various and latest physical or biological theories of Buffon, Diderot, Maupertuis. The author devotes many pages to recounting how the life of the "cacouacs" is built upon the interpretation of nature thus obtained, with a ritual to match. The telling point of Moreau's satire is the portrayal of sci-

---

correspondance philosophique, historique & critique, nouv. éd., La Haye, 1738, II, 340.

entific naturalism as a rival of the traditional theology, indeed to the point of having taken over, by its multiple functions, the authority of established religion for its adherents.[46]

In view of all the foregoing evidence, the philosophes in general, and the materialists in particular, could have found ample means to substantiate their ideas, had they so wished, by citing the appropriate aspects of Cartesianism from which these had sprung. It is surprising, at first, that this did not happen. Their allusions to Descartes, often deliberately vague in meaning, remained for the most part on a restrained, stereotyped level, and in themselves would barely provide a preface to his influence on the Enlightenment's naturalistic phase. Usually, Descartes was singled out as the destroyer of Scholasticism, the founder of modern philosophy, and the inventor of a new rational method in the use of which the philosophes described themselves as the real masters. The evaluation of Descartes's role in the inception of *lumières* seldom went beyond this stage of generality and banal utterance. Before attempting, however, to restore here the major part of what the philosophes omitted from the sum total of their obligations to Cartesian thought, one wonders why such a contribution should have been concealed or minimized. Indeed, Diderot, La Mettrie, and D'Holbach had good reasons for not placing their doctrine plainly under the aegis of Descartes; but these reasons were mainly of a tactical sort, and are not to be taken at face value in the reconstruction of ideological ties.

It is known, concerning the evolution of Descartes's ideas in France, that "inconsistency made him the source of two important but divergent schools of philosophy."[47] One must add, however, that whereas one school—the idealism of Malebranche and his followers—enjoyed a kind of official status

46 J. N. Moreau, *Nouveau mémoire pour servir à l'histoire des cacouacs*, Amsterdam, 1757, esp. pp. 34-64.

47 Russell, *History of Western Philosophy*, p. 591.

and allied itself with theological authority, the other had taken shape clandestinely as an "illegitimate" offspring of Cartesianism. As a result, the philosophes found themselves from the start in a false position vis-à-vis Descartes. Moreover, this situation was aggravated when, during the eighteenth century, those who most solemnly proclaimed themselves Cartesians happened also to be, for reasons which will be stated, enemies of the Enlightenment. Inasmuch as the innatist metaphysics of official Cartesianism was something that the philosophes, as diligent students of Lockean epistemology, rejected and combated, they understandably refused in good faith to style themselves Cartesians. For practical purposes, too, Diderot and his associates were reluctant to make Cartesianism openly a support for their opinions, since that would merely have embroiled them in futile polemics with their opponents about the valid sense of the designation.

The background of this ironic situation, owing to which the heirs of Cartesian reason were often to turn that destructive weapon on Descartes's philosophy, deserves briefly to be sketched. By the end of the seventeenth century, Cartesianism in France, with the aid of diverse accommodations, was well on the way to reconciliation with Catholic dogma. Much in its author's own attitudes and pronouncements on matters of faith had encouraged, or so it seemed, this "christianizing" tendency. Cartesian thought, though appearing decidedly agnostic in spirit to many of its contemporaries, and despite its presence on the Index—*donec corrigatur*—since 1663, had nonetheless carefully avoided an overt break with orthodoxy. The philosopher's "prudence," in this regard, has become classic. He had first obtained for his general enterprise the sympathetic approval of the Cardinal de Bérulle, and of the *Oratoire* founded by that pious prelate. Against anything that might savor of heresy in his writings, he had sought advice and corrections from a cleric such as Mersenne, who shared happily the same mechanistic view of nature as himself. Descartes had courted the favor, albeit in this case unsuccessfully, of the Jesuits. He had repeatedly professed a full submission to all articles of faith, and had desisted from any test of methodic doubt in the domain of revelation. Even his personal life had shown no signs of incredulity; he was supposed to have made the pilgrimage to Notre-Dame de Lorette out of gratitude for discovering the principles of his philosophy. The instances of Descartes's religious probity may be multiplied many times over. All in all, Baillet's standard *Vie de Monsieur Des-Cartes* had succeeded, without too much distortion of fact, in giving the portrait of a perfectly correct respecter of the Catholic creed and its practices.

If Descartes's works had treated, while remaining indifferent to questions purely of faith, the supreme problem of God's existence and attributes by demonstrable reason alone (in which some saw danger), the intention had been, commendably, to confound libertines and atheists by arguments suited to them. Even if, behind a pious exterior, Descartes had sought *cynically*, by propping his mechanistic, unprovidential physics with a spiritualist doctrine of Augustinian character, to neutralize the Church's reaction to his anti-Scholasticism,[48] this could not altogether remove for the philosophes the stumbling-block of his metaphysics. Descartes's thinking, whatever its innuendoes might be, had failed to exhibit that almost axiomatic opposition of reason to faith, and of the natural to the supernatural, which underlay the naturalism of Diderot's period.[49] All of these circumstances help to clarify the re-

[48] Etienne Gilson has found, on a given topic, that the elaborate theological apparatus present in Cartesian philosophy is subservient to the aim of rendering the physics acceptable. Cf., *La Liberté chez Descartes et la théologie*, Paris, 1913, esp. pp. 437ff.

[49] Henri Gouhier has concluded that neither Descartes's ideas, nor his conduct, exhibited a conscious inner conflict between religion and science. Cf., *La Pensée religieuse de Descartes*, Paris, 1924, *passim*. But the positive results of his analysis bear upon the estimate, not so much of the philosophes' actual debt to Descartes, as of their inability to acknowledge it properly. M. Gouhier states, moreover, that Descartes's attitude towards the relation between science and theology was essen-

serve with which the philosophes, and particularly the materialists, regarded Cartesian philosophy. Diderot might privately have agreed with D'Alembert that Descartes was "the first of the rebels."[50] But to make of him a philosophe *avant la lettre* publicly would have presented too many difficulties. The sifting of the revolutionary from the merely traditional elements in Cartesianism would have taken the philosophes far off their course. Theirs was principally an effort, not of philosophical exegesis, but of practical reform. Descartes's system was for the most part abandoned as unsalvageable.

In the eighteenth century, Cartesian thought came largely to be associated, by a historical process that ought to be outlined, with the forces of intolerance and bigotry. Descartes himself had merely affirmed a belief in the compatibility of the established religion and his philosophy; certain of his followers went much further in exploiting the latter's apologetic possibilities. The details of this general development are too well known to require any but the briefest mention here.[51] An early disciple such as Clerselier had tried to harmonize Cartesian ideas, then under attack from various quarters in the Church, with essential points of dogma. Rohault, interested mainly in propagating his master's physics, had deemed it wise, in the *Entretiens de philosophie* (1671), first to put the definition of matter as extension in agreement with the Eucharistic mystery. This last turned

tially "une vue thomiste," i.e. harmonious separation; pp. 272ff. The weakness of this, despite the documentation available from Descartes himself, is that from the seventeenth century down to the present time, Thomists themselves have stressed the basic unsuitability of Cartesian science, unlike the Scholastic, to lead the mind towards a contemplation of God. M. Gouhier undervalues, in our opinion, historical testimony of the non-Christian, indeed anti-Christian, character of Descartes's physics; pp. 294-300.

50 D'Alembert, "Discours préliminaire," *Œuvres*, I, 67.

51 Bouillier's *Histoire de la philosophie cartésienne*, although often tendentious and inclined to overstatement, remains still the definitive work to consult. See also, Joseph Souilhé, *La Philosophie chrétienne de Descartes à nos jours*, 1934.

out to be a specially delicate problem; having been dispelled by Descartes himself with little less than legerdemain, it was destined to try the ingenuity of certain of his adherents well into the next century. The case is instructive; but more important than such isolated issues and dubious accords was, for example, the ties that had early formed between Cartesianism and Jansenism. Through the efforts of Arnauld, the former came to be fused, and in some ways confused, with Augustinism, with which of course it had much in common.[52] When Diderot was to engage in debate with the Jansenist Bishop of Auxerre in defense of the abbé de Prades's scandalous Sorbonne thesis, he was to encounter in his path this particular species of Cartesianism.[53] But the new philosophy had enjoyed a more reputable entry into the Church than that through the ill-starred Port-Royal. As Fontenelle remarked, Malebranche did with the system of Descartes what Thomas Aquinas had done with that of Aristotle.[54] With the materials left by his predecessor, Malebranche elaborated the subtlest and profoundest apologetics of the age, converting innate ideas into a Christian-Platonic "vision in God," and mechanistic dualism into a providential theory of occasional causes. So considerable was the success of this that, among other effects, it seems to have convinced some of the philosophes that occasionalism was explicitly part of Descartes's system—an event which did not improve his reputation in their eyes.[55] In the alliance between Catholic dogma and Cartesianism, the latter had significantly found points of contact with a variety of religious attitudes. A champion of the Church such as Bossuet, while sensing that Descartes's thought might some day present a grave challenge to faith,

52 Bouillier, *op. cit.*, II, 156-177.

53 Diderot, I, 450-451.

54 Bernard Le Bovier de Fontenelle, *Œuvres*, nouv. éd., Paris, 1766, V, 398.

55 Fontenelle's *Doutes sur le système des causes occasionnelles* was directed against both Descartes and Malebranche. Voltaire, likewise, classed the two together on the same point: *Œuvres*, XXII, 424.

was nonetheless willing, by a curious eclecticism, to combine certain of its features with Thomistic tradition, as seen in the dualist metaphysics of his *Connaissance de Dieu et de soi-même.*[56] If with Bossuet Cartesianism had reached the threshold of officialdom in France, with a quite different interpretation, that of Fénelon, it became merged also with pietistic tendencies. In the *Traité de l'existence de Dieu* and the *Lettres sur la religion et la métaphysique*, Fénelon gave to Descartes's ontological proof of God's existence an equal, if not superior, weight by comparison with the cosmological argument of St. Thomas.[57] It was remarkable that Arnauld, Malebranche, Bossuet, Fénelon were all engaged, despite serious differences with one another, in the common task of assimilating Descartes to the rigid position of the Church.

Owing to such efforts, which had their effect, the eighteenth century was made to perceive the complex of Cartesian ideas through a kind of incense-laden, apologetic mist. With the conversion of Cartesianism into an arm with which authoritarians hoped to frustrate the goals of Enlightenment, it became difficult for the philosophes to deal objectively and dispassionately with notions originating in the system of Descartes. The *Méditations métaphysiques*, with its proof of God's existence, its substantial distinction of a spiritual realm, and its invitation to transmute innatism into revelation, had undoubtedly lent itself well to anti-philosophe, and specially to anti-materialist criticism. But the career of Cartesian physics, by contrast with the metaphysics, became involved mostly in paradox. When the Cardinal de Polignac said of his *Anti-Lucretius*, perhaps the best-known refutation of materialism in the epoch of *lumières*, that it was Cartesian in inspiration, he spoke only a half-truth. In reality, he made liberal use of Descartes's metaphysics; but the teleological argumentation that was the pith of the Cardinal's thesis, far from being a consequence of Cartesian science, was diamet-

[56] Cf., Bouillier, *op. cit.*, II, 227-263. [57] *ibid.*, 264-304.

rically at odds with it. Curiously enough, too, the most ardent partisans at the time of the *tourbillons* and related hypotheses, such as Père Castel, Père Regnault, Charles-Hercule Keranflech, Père Paulian, and others, were also hostile to the philosophes. But Descartes's physics—"cette Physique terrible, qui veut établir la Foi en la ruinant"[58]—proved to be not simply indifferent to religion intrinsically, but, as will appear, conduced irresistibly to naturalism.

How did the defense of vortex physics and the opposition to Enlightenment happen, then, to coincide among Diderot's contemporaries? The explanation of this anomaly is perhaps to be looked for in the apologetic phase of Cartesianism. Descartes had already been so completely assimilated (or so it was imagined) to the side of orthodoxy, that the cause of any specifically Cartesian doctrine—excepting the method of universal doubt, which was left to the infidel—was assumed uncritically by many spokesmen for the Church to be inseparable from the cause of Christianity itself. The case of the Jesuits is illuminating. The Society of Jesus had of course been from the start, and was still in the eighteenth century, the implacable foe of Cartesianism, and the upholder of Scholasticism. Yet even the Jesuits acquiesced in the end to the annexation of Cartesian thought to the Catholic viewpoint. The historian of this particular event has claimed that the Jesuits, like others, found in Descartes a powerful ally in the struggle against the deistic, materialistic, and atheistic tendencies of the Encyclopedist group.[59]

The counterpart of all this was that for the philosophes it became preferable to trace their philosophical or scientific opinions to almost anyone but Descartes: Bacon, Newton,

58 Montfaucon de Villars, *La Suite du comte de Gabalis, ou Nouveaux entretiens sur les sciences secretes, touchant la nouvelle philosophie*, Amsterdam [1715], p. 47.

59 Gaston Sortais, "Le Cartésianisme chez les Jésuites français au XVIIe et XVIIIe siècle," *Archives de Philosophie*, Vol. VI, cahier iii (1929), p. 92.

Lucretius, Epicurus, Gassendi—even when very likely the first-mentioned had wrought the immediate and lasting effect on their thinking. It is disconcerting that Descartes's name should have acquired, during the Enlightenment, an undeniable propaganda value which served, only too often, to obfuscate the actual relationship of his natural science to the materialist ideology. There was more than one sense to D'Alembert's quip about Descartes: "Above all, let us not confuse his cause with that of his sectarians,"[60] even if the philosophes were inclined to ignore all but its most commonplace meaning. The chapters following will seek to get around and beneath this artificial surface-pattern, according to which Descartes was bandied back and forth, with unusual facility but disappointing superficiality, by both philosophes and traditionalists. Only in this manner can it be understood how the great philosopher's ideas, despite misleading appearances, acted powerfully upon the intellectual issues that fashioned the scientific naturalism of Diderot, La Mettrie, Buffon and D'Holbach.

In conclusion, a brief comment would not be out of place concerning the extent of the philosophes', and particularly Diderot's familiarity with Cartesian philosophy. At their century's dawning, it must be noted that the most forceful, compelling thought-complex, in France at least, was Cartesianism, not only by its offer of rationalist criteria for the treatment of various questions, but more specially by its general orientation in the natural sciences. Until about 1735 Cartesian physics, notwithstanding the gradual infiltration of Newton's theories and experimentalist tenets, remained in authority on a most pervasive scale. It was in those years, when Cartesianism was the official position of French science, that the future exponents of scientific naturalism, Diderot, La Mettrie, Buffon, D'Holbach, received through formal schooling their initial intellectual formation. At the time, the

60 D'Alembert, "Discours préliminaire," I, 74.

Académie des sciences, as well as almost all learned societies in France, was permeated not simply with the diverse physical and biological doctrines, but more significantly, with the conception of science proper to Cartesian tradition. It was not until the 1730's that this hegemony of Cartesianism was questioned in a fundamental sense, and began at last to disappear. But the decline of Cartesianism was not exactly a rout; there is reason to believe rather that it was an orderly, even if inexorable, retreat. It left behind many all-embracing vestiges, which proved all the more effectual for being too "intangible" to be challenged by the philosophes: in a way, Cartesianism triumphed even in defeat. In 1765, Thomas's *Eloge* observed: "Time has destroyed the opinions of Descartes, but his glory subsists. He is like those dethroned kings who, even amid the ruins of their empire, seem born to command men. ... There are sterile truths, still-born as it were, that advance nothing in the study of nature; there are the errors of great men which become pregnant with truths."[61] The materialists had had a Cartesian introduction to the study of nature; it will be seen that the victory of Newton's experimental physics achieved in their lifetime was, in effect, a superposition of new theories and attitudes on the substratum of Cartesianism which remained.

In the case of Diderot specially, it is probable that the early acquaintance with Descartes was a thorough one. It would be pertinent to ask what place the teaching of Descartes's ideas might have had in the future encyclopedist's formal education. One knows, from the biographical sketch by his daughter, Mme de Vandeul, that Diderot completed his schooling in the liberal arts at the Collège d'Harcourt. Since he was born in 1713, his attendance there could not very well have been before the end of the 1720's. Some years prior, corresponding with the libertinism of the Regency, the University of Paris had reversed its timeworn policy towards the

[61] Thomas, *Eloge de Descartes*, pp. 4-5.

Cartesian system. Whereas the teaching of it had up until then been forbidden, Descartes's works were suddenly added to the curriculum to help combat the Regency's vogue of materialism and atheism.[62] Rollin's *Traité des Etudes*, of 1725, gives valuable information about the extent to which Cartesian philosophy had, soon after this change of policy, become an integral part of the program of studies in the various *collèges* belonging to the University of Paris. The standard technique of instruction in philosophy, according to Rollin, consisted in having the students read for themselves, or the best among them to the rest of the class, portions of such works as Descartes's *Méditations métaphysiques* and *Principes de la philosophie*, or Malebranche's *Recherche de la vérité*. After difficulties in the text had been resolved by the professor, the students were required to give a précis of the material covered, being attentive to logical order in their method of exposition.[63]

At the Collège d'Harcourt, attended by Diderot, the prevalence of Cartesian thought must have been even greater than elsewhere. From 1713 to 1730, the Collège d'Harcourt was administered by Guillaume Dagoumer, a distinguished professor of philosophy who, for the major part of his career, had championed the introduction of Cartesianism into the schools.[64] Rollin's *Traité des Etudes* finally vindicated and gave official sanction to the lifelong efforts of Dagoumer.

[62] Charles Jourdain, *Histoire de l'Université de Paris, au XVIIe et au XVIIIe siècle*, Paris, 1888, II, 176.

[63] Charles Rollin, *Œuvres complètes*, éd. Letronne, Paris, 1821-1825, XXVIII, 191-192. D'Alembert offers additional evidence of the important place of Descartes in the curriculum of the schools at the time, saying of his own education at the Collège Mazarin between 1730 and 1735: "Son professeur de philosophie . . . cartésien à outrance, ne lui apprit autre chose pendant deux ans, que la prémotion physique, les idées innées et les tourbillons." "Memoire de D'Alembert par lui-même," *Œuvres*, I, 1-2.

[64] Cf., H. L. Bouquet, *l'Ancien Collège d'Harcourt et le Lycée Saint-Louis*, Paris, 1891, pp. 334-335, 348. Dagoumer was outstanding enough to have been appointed Rector of the University on two occasions, from 1711 to 1713 and from 1723 to 1725.

There is good reason to believe, consequently, that when Diderot, shortly after, entered the Collège d'Harcourt, the dominant philosophy being taught there was Descartes's, with which the future encyclopedist was thus made fully conversant. Moreover, his broad reading of philosophical literature for the *Encyclopédie* allowed him, later on, to judge the original import of Descartes's system against the welter of doctrines, both new and old. Cartesianism had been added to the curriculum of the schools to discourage materialist attitudes. It is quite possible that the ideas and methods of Descartes, insofar as they proposed to interpret the physical universe, defeated this purpose and, in the minds of a few students, such as Diderot, laid the basis instead for a subsequent naturalism. It remains to be examined just how this might have occurred.

CHAPTER II

FROM DESCARTES'S *MONDE* TO THE WORLDS OF DIDEROT AND MATERIALIST SCIENCE

"Donnez-moi de la matière et du mouvement, et je ferai un monde."

—Voltaire on Descartes's physics, *Œuvres*, xxii, 404.

CHAPTER II

FROM DESCARTES'S *MONDE* TO THE WORLDS OF DIDEROT AND MATERIALIST SCIENCE

DESCARTES had written to Mersenne: "If anyone could know perfectly what are the small parts composing all bodies, he would know perfectly the whole of Nature."[1] Cartesian physics was in this particular sense (for the term has other connotations) corpuscular. It reduced natural science exclusively to a consideration of the components of matter: their sizes, shapes, interactions. Both final or "intelligent" causes and all varieties of occult or supernatural agents were ruled out by Descartes as factors in the explanation of physical events and processes. His doctrine of innate ideas, or those *a priori* notions thought valid for establishing the foundations of physics independently of the aid of the senses, did not conflict with this corpuscular conception. Indeed, innatism served to corroborate the basic assumption that natural philosophy dealt ultimately with small particles of matter in motion, and with nothing else. The author of the *Principes de la philosophie* had asserted: "I considered in general all the clear and distinct ideas that can be in our understanding concerning material things, and not having found any but those of figure, size, and motion, and the rules according to which these can be diversified by one another, which rules are the principles of geometry and mechanics, I judged that all the knowledge men could have of nature had necessarily to be derived from this only; because all the other notions we have of sensible things, being confused and obscure, cannot serve to give knowledge of anything outside us."[2] These words were aimed specifically at

1 Letter dated 9 Feb. 1639, in Descartes, II, 497.
2 *ibid.*, IX, 321.

the science of the Schoolmen, which resorted to such principles as "intentional species," "occult forces," "sensible qualities," "final causes," and the like. But the Cartesian dictum was no less capable of serving, into the future, as motivation for the trend to scientific naturalism that made rapid headway in eighteenth-century France.

By reference to his program in natural science, Descartes's metaphysics had been specially well-chosen. The ontological proof of God's existence and attributes had in effect assured the highest degree of autonomy from theology to Descartes's philosophy of nature. It is permissible to think that his predilection for an Augustinian metaphysics was due precisely to this unprecedented, unlimited measure of freedom vouchsafed to his physics. By resolving certain primary questions of a theological character entirely apart from the order of physical phenomena, Descartes had cleared the path, conversely, for treating of nature wholly apart from theology. With a method presumably advantageous to the demands both of theologians and of the new science, he had elaborated the theology of the *Méditations métaphysiques* by abstracting from the domain of sense-experience and relying, instead, uniquely on the logical consequences of certain innate, self-evident ideas concerning God, infinity, and perfection. A principal result of this Cartesian metaphysical scheme—the result of interest to the present study—is that it rendered the sphere of mechanical contingency and physical fact, that is, Nature seen as the proper object of science, unrelated in any fundamental sense to problems of a theological or religious kind.[8] This view of Nature, with certain principles and hy-

[8] It is not meant that the *Méditations métaphysiques* was merely a "front" which permitted the publication of an anti-scholastic physics under the shield of a seemingly orthodox metaphysics. On the contrary, Descartes's metaphysics was organically indispensable to the success of his natural science. Its role was precisely that of settling the questions of chief concern to theology in a kind of limbo of pure reason, in order that physical inquiry, in its actual contact with phenomena, might all the more logically divest itself of theologically prescribed notions.

potheses drawn from Descartes's physics, was to engender in due time the materialist ideology of Diderot and his colleagues.

Having definitively severed physics from theology, Descartes felt justified to give, in the *Monde ou Traité de la lumière* and in the *Principes*, his own non-theological version of Creation. The forming of Descartes's universe, which served as general introduction to his system of nature, made the actual state of the world deducible, in all its broad as well as detailed features, from the initial properties of matter and the laws of motion. In the *tourbillon* hypothesis, which was a concrete instance of how this vast cosmogonic deduction might be carried out, the point to be stressed is that only two ultimate principles of physical causation were recognized: matter and motion. What transcended these was, properly speaking, no longer physics, but either metaphysics or theology. And as a physicist, Descartes stated clearly his determination to reproduce theoretically, from matter in motion alone, a world in which, "although God should give it no order or proportion . . . one will be able to see . . . all the things, both general and particular, that appear in the real world."[4] True, the philosopher had elsewhere claimed for the laws of motion a logical deducibility from the definition of God. But judged against the procedure embodied in the *Monde* and *Principes*, the connection between Deity and mechanical law was, if not actually negated, at least so closely restricted in meaning as to become irrelevant to the general problem of natural science.

The discrepancy between the assumed theological origin of the laws of motion and the specific use made of them in the non-theological construct of the *Monde*, was to be exploited historically to the profit of scientific naturalism. This was confirmed quite explicitly, in 1754, by Denesle's *Examen du matérialisme*, which contended that Descartes's rejection of

[4] Descartes, xi, 34-35.

finalism could only have resulted in the type of naturalistic science which, at the time, was steadily gaining favor everywhere. Denesle said of Cartesian physics: "The philosopher who with matter and motion promised to make a world like the actual one, was not careful what he was saying . . . for he should have realized it would be in vain that movement was imprinted on matter by an infinite power, if a supreme Intelligence had not given to this movement directions that conformed with his views."[5] And in 1761 the abbé Gabriel Gauchat, striving to refute the materialism whose advances the philosophes were busily consolidating, put the blame for the entire development squarely on certain implications of Descartes's above-quoted statement of objectives in the *Traité du Monde*. Some physicists had apparently, despite their intentions, played into the hands of materialism: "In granting to motion alone responsibility for the original arrangement of the globe, they lend arms against their wish to those who seek to attribute everything to matter." Gauchat's protest was about "certain principles of physics, which are capable of giving imaginary powers to matter and of abrogating unjustly those of the Creator."[6] The pages following amplified on how such a way of thinking had been prompted above all by Descartes's natural philosophy. To know fully what the abbé Gauchat had in mind, however, one must retrace with some care the historical current which had prepared Cartesian ideas for the use that Diderot's generation made of them. Diderot, Buffon, D'Holbach did not submit to this influence in a temporal void. A continuous and persistent stream of interpretation had already extracted from Descartes's sidereal physics its naturalistic implications, and composed of these a definite ideological atmosphere, when Diderot and his contemporaries made their debut.[7]

[5] Denesle, *Examen du matérialisme*, Paris, 1754, I, 43-44.

[6] Gabriel Gauchat, *Lettres critiques, ou analyse et réfutation de divers écrits modernes contre la Religion*, Paris, Tome XV (1761), p. 265.

[7] Pierre Busco, in *Les Cosmogonies modernes*, Paris, 1924, esp. pp.

The naturalism inherent in Descartes's science had, from the outset, been noticed and condemned. Voëtius zealously opposed the new physics, in 1641, on the grounds that its fundamental divergence from Scholastic concepts rendered it incompatible with the cosmological teaching of Moses. Moreover, the rigorous laws and broad potentialities ascribed by Descartes to mechanism as such seemed to Voëtius, as compared with the Schoolmen's reliance on a scheme of "intelligent" causes behind natural events, to be pernicious for the belief in a rational soul and in the mysteries of the Christian faith.[8]

Across the Channel, the foreseeable outcome of Cartesian science exercised the critical acumen of Henry More. Having at first hailed with enthusiasm the apparent solidity of the new spiritualism, the Cambridge Platonist came, in maturer years, to recognize the ability of Descartes's physics to nullify the merits of the *Méditations métaphysiques*. The *Enchiridion Metaphysicum* of 1671, a lengthy refutation in form of the Cartesian system, left no doubt about its motivation: "No greater blow and wound can be inflicted on the most essential part of Religion than by presuming the possible explanation of *all Phenomena clearly through their causes* (the structure of plants and animals not excepted). As if this material world, provided merely that one supposed as much motion communicated to matter as indeed one finds in it, could have engendered itself! Yet such is the Cartesian hypothesis."[9] More complained that Descartes had admitted the

13-48, has showed how the salient features of the *Monde* laid the basis for the development that scientific cosmogony received later in the efforts of Buffon, Kant, Laplace, and their successors. Our purpose is to study, not Descartes's role in the history of cosmogonical inquiry as such, but the contribution of the *Monde*, up until 1750 or so, to the growth of a materialist ideology which, in turn, the philosophes applied to various problems in natural philosophy, the theory of science, ethics, and religion.

8 Baillet, *Vie de Monsieur Des-Cartes*, II, 146.

9 Henry More, *Enchiridion Metaphysicum: sive, de rebus incorporeis succincta & luculenta dissertatio*, London, 1671, "Ad lectorem praefatio," §4 (unpaginated).

existence of a spiritual principle only at the drastic price of expunging it completely from the realm of nature.[10] Judging dualist metaphysics a feeble preventive against the materialism (or atheism) to which such mechanistic philosophy must lead, the Englishman predicted remarkably well the impact of Cartesianism on certain philosophes.[11]

The misgivings of Descartes's own contemporaries were to be revoiced and confirmed during the Enlightenment when, in Brucker's *Historia critica philosophiae*, it was pointed out how dangerous had been the unlimited activity of mechanism posited by Cartesian thought. Brucker revealed, despite a reserve induced by disapproval, the result of Descartes's habitual recourse to non-rational, efficient causes: ". . . since things are so, and it is desirable to rescue Descartes from the ranks of clandestine atheists, we strongly deplore that he was slave to a ruinous objective and guilty of a certain degree of carelessness. Not only are both these traits applicable by philosophers of profane mind to the defense of impiety, but are actually so applied by them to some extent."[12]

Returning to seventeenth-century France, the keen intelligence of Pascal had early seized the radical possibilities of the physics of Descartes who, observed the *Pensées*, "would have wished . . . to be able to do without God, but could not avoid according Him a flip of the finger in order to set the world in motion; after that, he has no further need of the Deity."[13] Like Descartes, Pascal was a student of mechanistic

[10] *ibid.*, "O duram rerum incorporearum conditionem, (vel potiùs ridiculam & impossibilem,) quae eâ solâ lege in rerum universitatem à *Cartesio* admittuntur, ut nusquam in universa rerum Natura exsistant!"

[11] *ibid.*, §6: "Verùm enimvero, ut res se nunc habent, si Philosophiae Cartesianae, quà Physicae quà Metaphysicae, staretur, equidem horreo dicere in quanta proclivitate & quàm periculoso in Atheismum praecipitio mortalium animi constituerentur."

[12] Johann Jakob Brucker, *Historia critica philosophiae a mundi incunabilis ad nostram usque aetatem deducta*, Lipsiae, 1742-1744, T. IV, part ii, p. 334. Diderot used this work extensively as a source for his numerous *Encyclopédie* articles pertaining to the history of philosophy.

[13] Blaise Pascal, *Pensées et Opuscules*, éd. Brunschvicg, Paris (n.d.), pp. 360-361.

science; but he perceived the risk of setting down *too* much to the credit of matter in motion, as had done the *Monde*. Alluding still to Descartes he added: "It ought to be said in a general way, 'all this occurs by configuration and by movement,' for that is true. But to tell by what specific motions and configurations, and so to compose the mechanism, would be ridiculous; for that is useless, uncertain and difficult." Pascal had deftly put his finger on the exact Cartesian theme that was to transform itself into the naturalism of the next century. By comparison, quite misguided and futile were the rather quixotic attempts of those who, from Amerpoel's *Cartesius Mosaïzans* (1669) to Keranflech's *Observations sur le cartésianisme moderne* (1774), sought to prove to the contrary that Descartes's account of Creation tallied with *Genesis*. The verdict of history was to be against them, and was to bear out the suspicions of Pascal.

There is evidence, towards 1700, that Cartesian science had already begun to support freethinking tendencies. An indication of this may be seen, for example, in Montfaucon de Villars' *La Suite du comte de Gabalis*, composed much earlier than its publication in 1708. While generally satirizing Descartes's system, de Villars' work is important for the attention it called to the relationship between certain Cartesian aims and the pantheist speculations of the Renaissance cosmogonist, Giordano Bruno. From this analysis, the *Suite du comte de Gabalis* drew conclusions about Descartes's physics which are a fair sample of the intellectual attitudes that culminated in mid-eighteenth-century materialism. Concerning the origin of the universe depicted by vortex-theory, de Villars commented:

". . . it is solidly founded on demonstrations of Mechanics which are so beautiful, natural, and necessary, that to be quite frank, it is altogether useless to suppose that God had anything at all to do with this subject; with the production of the sun, of the light assumedly surrounding it, and of all the rest. If Scripture did not teach us that God had labored seven days to make the world . . . we would have spared Him all that effort, provided only that he created for us, as He has done, matter divisible to infinity, and consisting of little corpuscles shaped like dice and screws."[14]

In the particular evolution of ideas here under study, it should be stressed that the broad and potent spread of Cartesianism, in a major phase, absorbed into itself and thereby revitalized the doctrine of classical Epicureanism. It was the merger of certain Cartesian principles with analogous notions deriving from ancient atomism that, in effect, exerted its combined force on Diderot's circle. Despite Descartes's famous polemic with Gassendi dealing, for the most part, with the theory of knowledge and the problem of metaphysical dualism, there had perhaps been, with regard specifically to natural philosophy, more in common between the two thinkers than either had cared to admit, or so at least the history of ideas was to suggest. Significantly, in Descartes's own time his opponent Plempius published, in 1654, a series of letters whose main theme of criticism was that the new physics rested on the same foundations as that of Democritus and Epicurus, "however much expurgated and embellished it might have been by René Descartes."[15] The materialist intimations of Cartesian science were, furthermore, notably strengthened by the efforts of certain exponents who, like Rohault and Régis, divorced overtly its physical theories from antecedent metaphysical conditions. Montfaucon de Villars was able, before long, to put in the mouth of a disciple of Descartes a discourse expressing the basic affinity, as then perceived, between Cartesian and Epicurean thought:

"There is no divergence at all . . . between that doctrine

[14] Montfaucon de Villars, *Suite du comte de Gabalis*, p. 35.
[15] Georges Monchamp, *Histoire du Cartésianisme en Belgique*, Bruxelles, 1886, pp. 247ff., 262.

[the Epicurean] and ours as regards essential matters. For . . . what difference does it make to religion and faith whether the particles of matter are square or irregular, and whether they follow a diagonal-perpendicular, or a circular motion; provided that both of these occur necessarily and that a mechanical order results therefrom, without the need existing of recourse to a divinity or to anything which might be called spiritual?"[16]

Ironically enough, it would seem that Descartes's doctrine superseded Gassendi's as the progenitor of French naturalism. There are, however, several definite reasons for such an event which, otherwise, might well appear paradoxical. Whereas the Epicureanism revived by Gassendi and harmonized with the position of the Church had not instituted any *scientific* school in the proper sense, it will be observed that the materialism of Diderot's group found its motivation and purpose, above all, in the progress of the sciences of nature and in the requisite affirmation of their legitimate authority. In the interim from 1680 to 1750, it would be difficult to describe as specifically Gassendist, and not indebted for its general principles, hypotheses and methods to Descartes, Newton, or to both, a single serious work of physics. Whatever Epicureanism was present in such questions during that long period, then, was owing primarily to the fresh meanings given by Descartes and Newton, as well as by their followers, to the classical notions of "void," "corpuscle," "subtle matter," "vortex," and the like. It was, moreover, Cartesian tradition which, by its dominant influence in France up until the 1730's, actually furnished the ingredients essential to the philosophes' scientific naturalism.

The early absorption of Gassendi's teaching into the purview of Cartesian natural philosophy is concretely perceived in a figure such as Cyrano de Bergerac. The *Etats et Empires*

[16] Montfaucon de Villars, *op. cit.*, pp. 56-57.

de la Lune, completed towards 1648, had reflected distinctly Cyrano's course of instruction under Gassendi in 1641. Afterwards, however, the celebrated libertine came under the progressive sway of Descartes's physics, largely through his friendship with Rohault whom he met around 1648. In consequence, Cyrano's *Histoire des Etats et Empires du Soleil*, composed between 1650 and 1655, contrasted sharply with the previous work by finding its inspiration wholly in Descartes.[17] Written about the same time, Cyrano's *Fragment de Physique* was so faithful an outline of Cartesian theory that Rohault incorporated much of it into his own *Traité de Physique* of 1671.[18]

Cyrano's efforts in "fantastic science" literature stand at the head of a current of *libre pensée* which, by the time it reached the Enlightenment, was to have acquired a measure of seriousness. Meanwhile, the disciple of Gassendi had given to Epicureanism a basis in Cartesian physics. Cyrano's cosmic voyage in the *Etats du Soleil* unfolds against the background of the vortex universe's formation, which in turn will serve as general setting for a naturalistic view of things:

"I noticed . . . that all these worlds have other smaller worlds revolving around them. Dreaming since about the causes of the construction of this great Universe, I imagined that during the transformation of Chaos into order, once God had already created matter, similar bodies joined each other by that unknown principle of attraction [*amour*]. . . . Particles having certain forms assembled together to produce the atmosphere. Others whose figures made possible a circular motion, by uniting, composed those globes called stars, which

[17] Cyrano de Bergerac, *Les Œuvres libertines*, précédées d'une notice biographique par Frédéric Lachèvre, Paris, 1921, I, 2. The *Etats de la Lune* and *Etats du Soleil* were both published posthumously, in 1657 and 1662 respectively. After a long period of relative indifference to Cyrano, the reediting of his *Œuvres* in 1741 suggests a renewal of interest owing to the recrudescence of materialist ideas.

[18] *ibid.*, xcixn.

. . . by their tendency to pirouette on their poles, have not only necessarily assumed a round mass such as we observe, but have also determined the revolution of lesser orbs situated in the sphere of their action. That is why Mercury, Venus, the Earth, Mars, Jupiter, and Saturn have been compelled to spin and roll in unison around the Sun."[19]

Still adhering to Descartes, Cyrano sees this universe subject to constant vicissitude through mechanical process. He explains, like Descartes, even the phenomenon of sunspots in relation to the successive creation and destruction of worlds which, a century later, was to be an important corollary theme of French materialism:

"It is not unimaginable that formerly all these globes were Suns. . . . But these Suns, in the compass of time, have suffered so great a loss of light and heat by the continual emission of the corpuscles causing such phenomena, that they have become cold, dark and almost powerless pulps. We find even that sun-spots . . . increase in size from day to day. Now, who knows if these are not a crust forming on the Sun's surface from its mass that cools in proportion as light is lost; and if the Sun will not become, when all this highly mobile matter has abandoned it, an opaque globe like the Earth?"

Cyrano elsewhere remarks that the *tourbillon* cosmogony furnished a general foundation for the atomism of Epicurus.[20] As for Descartes's physics, "one ought to read it with the same respect due to the pronouncements of oracles."[21] The *Etats du Soleil* ends abruptly with the final, appropriate appearance on the scene of Descartes's ghost itself.

By the end of the seventeenth century, Cartesianism had assimilated to its own position many of the physical concepts of ancient Epicureanism renewed by Gassendi. One of Descartes's most hostile critics, the Jesuit Daniel, rather bitterly

19 *ibid.*, 127-128. 20 *ibid.*, 184. 21 *ibid.*, 188.

acknowledged this situation, at the time, in his popular satire on the *Monde* and its author: "You have even had the good fortune to efface in some manner all the recent philosophers who have appeared contemporaneously with you, and since. I've known of more than one adventurous thinker who, in the course of a dispute, has put M. Gassendi on the list of your disciples."[22] By contrast with this, the Epicurean school founded by Gassendi would seem to have acquired a peculiarly belletristic character. Diderot himself, in the *Encyclopédie* article devoted to "Epicuréisme," gave the genealogy of the spiritual descendants of Gassendi. Among them are to be found such famous *bels-esprits* as Saint-Evremond, Ninon de Lenclos, the abbé de Chaulieu, the comte de Grammont, and consonant names.[23] With a single exception—that of Fontenelle, to be considered presently—there are no scientists in Diderot's quite thorough enumeration of the Epicurean coterie. The moral and literary *volupté* that Gassendi, perhaps contrary to his expectations, appears to have inaugurated, undoubtedly made its contribution to scientific naturalism by fostering a vogue of freethinking to which La Mettrie, Diderot, D'Holbach were much indebted, and in which certain Cartesian ideas found conditions favorable to their growth. But the "system of nature" common to the philosophes just named, and which provided the theoretical underpinnings of their materialism, came from a source other than Gassendi's revival of classical Epicureanism.

As regards physical and biological doctrine, there is reason to believe that Descartes, perhaps more than Gassendi, improved upon and infused new vigor into the Epicurean tradition. Baillet offers a specially relevant confirmation of this. Descartes, it is said, had taken the trouble to correct those who likened his physics to Democritus'. In discussing the specific similarities and differences on that score between the two

22 Gabriel Daniel, *Voyage du Monde de Descartes*, nouv. éd., Paris, 1703, p. 257.

23 Diderot, XIV, 526-527.

philosophers, Baillet, following a passage in the *Principes*, concedes that the entire corpuscular basis of science was the same in Democritus and in Descartes. The divergences between them are then given as follows: Democritus had maintained the indivisibility of the ultimate components of matter, while Descartes upheld their (at least theoretical, even if not practical) divisibility; Democritus had attributed ponderability to the individual atom, whereas Descartes made each corpuscle ponderable only in relation to others in an interdependent whole; Democritus had asserted the reality of the vacuum, Descartes that of the *plenum*. These differentiae, important as they might be for framing particular hypotheses, are patently not of a kind that would render the purely corpuscular physics of Descartes any less "materialistic" than that of Democritus. Accordingly, Baillet comes to the last dissimilarity: Democritus differed from Descartes, "because he did not show the manner in which each thing could be produced by mechanical impact alone or by the concourse of corpuscles; or if he did indicate it in some cases, his explanations had no sequence or connection either among themselves or in relation to his principles, which is one of the chief differences between his philosophy and M. Descartes's, for the latter is entirely logical."[24] But if this concluding statement means anything, it means that Descartes was more thoroughly and consistently "Epicurean" than Democritus had ever succeeded in being. Through the medium of the Cartesian reform, in effect, the Epicurean natural philosophy may be said to have lost its elements of mythology and its subordination to a metaphysics built largely around the evasive concept of Chance; and instead, to have been integrated into the methodical, self-corrective search, typical of the modern era, for the fixed and determinable laws governing physical events. Descartes had made materialism, so to speak, "scientific." Cartesian science, based on a corpuscular and un-

[24] Baillet, *Vie de Monsieur Des-Cartes*, II, 227.

finalistic view of matter in motion, thus was able to convert to its standpoint the correlative parts of Gassendi's "physics." Epicureanism was thereby given a new career in which the names of Descartes, Gassendi, and their Graeco-Roman predecessors were often to occur in association. The historical background of the philosophes' scientific naturalism may be described as predominantly "Cartesio-Epicurean," with classical tradition supplying the generic but remote, and Cartesianism the specific and immediate, influence.

In the transmission of this thought-complex to the eighteenth century, most instructive is the role of Fontenelle, about whose success in propagating vortex-theory Montesquieu was to say in half-jocular fashion: "There are Cartesians who have read nothing but the *Worlds* of M. de Fontenelle."[25] Fontenelle figures on Diderot's list as an adept of "Epicuréisme"; but as a student of physics he was, during his long life that linked the philosophes' age with Descartes's, a steadfast Cartesian, vindicating as late as 1751 the vortices against attractionism. His *Dialogues sur la pluralité des mondes* (1686) offers early evidence of how Descartes's science nurtured naturalism, however restrained its expression had at the time to be. In the *tourbillon* astronomy that Fontenelle seeks to render intelligible to his "belle marquise," the conclusions drawn from their nocturnal lessons, suggesting already the future course of materialist opinion, are a logical extension of the *Monde's* theories. Each vortex, speculates Fontenelle, has by the laws of matter in motion produced its own world, or rather, worlds, since the various planets of a solar system, given certain climatic conditions, must be assumed inhabited like the earth. The notion of a pluralistic universe, absorbed into this agglomerate of neo-Cartesian physical doctrine, became simply an inference drawn from the constant ability of an indefinitely (or in-

[25] Montesquieu, *Œuvres*, VII, 82. Discourse read to the Bordeaux Academy, 15 Nov. 1725.

finitely) extended substance to engender, by its motions, a fittingly great number of worlds together with their inhabitants. Such a revaluation of an age-old theme invited difficulties with the Church concerning both the status of Adam as the forebear of mankind and the mission of Christ as its Redeemer. By contrast with the Middle Ages and Renaissance, which had often put forth the plurality of worlds and the infinity of the Creation as metaphysical implicates of the boundless power and goodness of God, the connotations of the idea had changed radically since Descartes's suggestion, in a letter to Chanut, that his cosmogony supported the belief that intelligent beings peopled other parts of the cosmos.[26] Nor did Fontenelle's concession that the denizens of other globes need not be human greatly alter the dangerous overtones of the *Dialogues*. Scientific naturalism, particularly in Buffon's *Théorie de la terre*, was to cite as a subsidiary argument the strong probability that, given the self-

[26] Descartes, v, 55; letter dated 6 June 1647. Arthur O. Lovejoy's *The Great Chain of Being*, Cambridge (Mass.), 1936, pp. 99-143, has traced, in relation to the maxim of plenitude, the career in sixteenth- and seventeenth-century Europe of the twin ideas of an infinite and a pluralistic cosmos. In the case of Descartes and Fontenelle, however, his analysis is somewhat misleading, for while the conceptions remain outwardly similar, their logical grounds have shifted. In Cartesian philosophy, the picture of a universe without fixed limits resulted, not from the principle of plenitude, but from the theory of the *plenum*; the latter, in turn, was deduced, not from the metaphysical rule of sufficient reason which postulated the realization in fact of all possibles, but from the definition of matter as *res extensa*. Likewise, for Descartes and his followers the plurality of worlds became, not a metaphysical necessity ensuing via the law of plenitude from the attributes of God, but a necessary consequence of the definition and properties of matter. These distinctions are indispensable to understanding the role played by the notion of an unlimited, pluralistic universe in the philosophes' naturalism. For instance, the same ideas expressed contemporaneously in England from the time of Henry More to Young's *Night Thoughts* (1745), and even later, were regarded as perfectly orthodox, while across the Channel they were unmistakable marks of freethinking. The explanation is simple. In England, the view of an infinite, pluralistic cosmos continued to be understood in conformity with the older theological argument for an infinite Creation by an infinite Deity. But interpreted in terms of Cartesian science, which predominated in France, the identical opinion could not but point to materialism.

determining virtues of matter and certain purely physical requirements for engendering and perpetuating organic life, other planets in the immensity of space were worlds more or less like our own. Presented wholly apart from metaphysical considerations, such speculation contributed to the rise of a consistent materialism by rendering irrevocable the opposition and rivalry between the traditional theology and the new science of nature.

Two basic factors, both integral to Cartesian philosophy, stand out in Fontenelle's envisagement of physical fact. First, finalism is excluded from his plan of exposition: "Our folly, peculiar to us humans, is the belief that the whole of nature, without exception, has been destined to our use."[27] Secondly, in searching out the causes behind the cosmic order, Fontenelle carefully avoids mention of a *deus ex machina*, but instead refers uniformly to the processes of a sort of demiurgic Nature that, unaided, has assumed the full responsibility for its production.[28]

From Descartes to Fontenelle, a decisive change has taken place in the evolution of Cartesianism. In Descartes's system, considered in its explicit meaning, physics had been subordinated logically to metaphysics and the laws of motion had been deduced, or so it was claimed, from the definition of God. Fontenelle's popularization emancipated itself from both metaphysics and theology. The laws of motion were conceived as properties of matter and made *per se* the starting-point of natural philosophy, without being further related to a cause outside and above nature. Such an adaptation of Descartes's science might appear at first glance to represent

27 Fontenelle, *Œuvres*, II, 23.

28 Fontenelle says of Nature, *ibid.*, 98: "Croirez-vous qu'après qu'elle a poussé ici sa fécondité jusqu'à l'excès, elle a été pour toutes les autres Planètes d'une stérilité à n'y rien produire de vivant?" Or, concerning the limitless extent of the pluralistic universe founded on the *tourbillons*, *ibid.*, 140: "La Nature n'a rien épargné en le produisant; elle a fait une profusion de richesses tout-à-fait digne d'elle."

a gross inconsistency, and to be completely at odds with his metaphysical position. Yet from the vantage-ground of intellectual history, this incipient naturalism was not altogether an illogical development. To an eighteenth-century mind such as Turgot's, for instance, the progressive separation of physics from metaphysics was something that had been initiated by Descartes. The *Histoire des progrès de l'esprit humain*, surveying the question broadly, declared: "How long a time was not required for the discovery that all sensible phenomena could be explained by means of figures and movements? Descartes was the first to have clearly perceived this truth. Until his arrival, physics had remained, for lack of this degree of analysis, more or less confused with metaphysics."[29] Moreover, in transit to the epoch of Enlightenment, Cartesian physics had frequently been abstracted from its fuller context of dualist metaphysics. In a sense, this was even a justifiable act. The link between Descartes's physics and metaphysics had, in fact, never been as indissoluble as its author had imagined and insisted. Recent examinations of the problem have tended to find that the basic principles of the Cartesian system of nature—for example, the laws of motion—do not derive absolutely, despite what Descartes himself asserted, from certain propositions concerning *res extensa*, the attributes of God, or innate ideas.[30] It was perhaps to be expected, then, that the process of history, by deepening this line of cleavage, would eventually have dissociated the physical from the metaphysical portion of the Cartesian philosophy, as exemplified by Fontenelle's *Pluralité des mondes*. Now, Descartes's physics itself, taken independently of the theological premises to which it had been somewhat arbitrarily subjected, proved capable of prompting and guiding materialist thought.

29 Anne-Robert-Jacques Turgot, *Œuvres*, 1808, II, 287.

30 See S. V. Keeling, "Cartesian Mechanism," *Philosophy*, Vol. 9 (Jan. 1934), pp. 51-66; also, Hyman Stock, *The Method of Descartes in the Natural Sciences*, New York, 1931, pp. 57-87.

A singular insight into such an ideological evolution may be derived from Bayle, who was early in life, as seen in his *Système de philosophie* summarizing lessons taught at Sédan and Rotterdam, an ardent professor of Cartesianism. But Bayle must soon have realized, in the face of concerted efforts to christianize what was orthodox in Descartes and to stamp out what was not, that the real and deep impact of the new doctrine would have to be a disguised, unofficial one. Like Fontenelle, Bayle may be regarded, in his own way, as a particularly important figure in the conveying of Cartesianism (albeit often in an indirect and surreptitious form) to the philosophes, on whom his influence is known to have been extremely strong. The *Dictionnaire historique et critique* sought, among other things, to point out the underlying continuity of physical principles from ancient Epicureanism to Descartes. Bayle's article "Démocrite" stated that Descartes's elimination of all secondary qualities from the domain of natural science coincided with the position of the Greek materialist, who had "maintained that there was no reality save for the atoms and void, and that all the rest was a subject merely for opinion. That is what the Cartesians say today concerning corporeal qualities such as color, odor, sound, taste, heat, cold; these are in their view only modifications of the soul."[81] In "Leucippe" an occasion offered for indicating, furthermore, the marked resemblance between the *tourbillon* universe and that of the founder of atomistic philosophy: "It would be impossible to deny that certain things in the Cartesian system are similar to the hypotheses of Leucippus."[82] Given the origination of vortex-theory by Leucippus, Descartes "has simply renewed old ideas" (and corrected, incidentally, his forerunner's erroneous notion of centrifugal force): "One finds . . . in the Leucippean system the seeds of that great law of mechanics which M. Descartes

81 Pierre Bayle, *Dictionnaire historique et critique,* Paris 1820, v, 459.
82 *ibid.,* IX, 196.

employs so efficaciously, namely, *that all bodies revolving around a center pull away from it as much as possible.*"[33] Bayle goes on to epitomize, in words that will remain valid a century later, the fundamental difference between Epicureanism and its counterparts in Gassendi and Descartes: "The moderns have distinguished more carefully; they reject the eternity of the atoms and their fortuitous motion, but in keeping, except for that, the hypothesis of Leucippus, they have built with it a most beautiful system. That is what Gassendi has done, who differs from Descartes concerning the principles of matter only by his retaining the vacuum."[34]

Elsewhere, Bayle comes closer still to defining the naturalistic orientation of Cartesian physics. Insofar as Descartes's formula of matter in motion had replaced the Aristotelian entelechy in the explanation of generative process, the *Monde* and *Principes* had reinstated the Epicurean contention, minimized by Gassendi, that all *development* in nature was to be looked for in mechanical, rather than intelligent, cause. Bayle's language, even, betrays here his attachment to the Cartesian school: ". . . if, in the seventeenth century, physics reappeared with a new lustre, that was owing only to the restoration of classical principles from which it had strayed. This was accomplished by the cultivation of evidence, and by the decision to exclude from the doctrine of generations that large number of entities about which our mind has no idea, and to think in terms of the figure, motion and situation of the particles of matter, all of which are things that one conceives clearly and distinctly."[35] In the article "Ovide" Bayle specifies, with minimum caution, the full measure of materialism inherent in Descartes's cosmogonic project, fusing this in turn with certain features of Gassendist science. It is stated that "nos nouveaux philosophes," having abjured the "qualities" and "faculties" of Peripateticism, are

33 *ibid.*, n. "B," 197. 34 *ibid.*, n. "D," 199.
35 *ibid.*, art. "Aristote," n. "M," II, 367.

left with the "general laws of motion, the principles of mechanics, the modifications of matter, and the figure, situation and arrangement of corpuscles" as the composite equivalent of the Aristotelian "active force" in nature. As regards the "four elements" common to both Peripatetic and Cartesian physics, Bayle's analysis augurs with surprising accuracy the kind of naturalism that the eighteenth century was to derive from a Cartesio-Gassendist complex of thought:

". . . these four elements, situated according to their natural lightness or heaviness, are a principle that suffices for all generations. The Cartesians, Gassendists, and other modern thinkers ought to maintain that the movement, situation and shape of the particles of matter are sufficient for the production of all natural effects, without excepting even the general arrangement that has placed the Earth, atmosphere, oceans, and stars where we see them. Thus, the true cause of the cosmos and of the phenomena occurring therein is in no manner different from the cause that has given motion to the components of matter, granting that it has simultaneously assigned to each atom a definite figure (as Gassendists hold), or has simply communicated to regularly cubic corpuscles an impulsion which, by the duration of a motion reducible to certain laws, would have them afterwards assume all sorts of shapes (such being the hypothesis of Cartesians). Both groups ought to agree, therefore, that if material substance was before the generation of the world such as Ovid has pretended, it would have been capable of drawing itself out of a state of chaos by its own powers, and of giving itself the form of a cosmos without the help of God."[86]

With the Ovidian cosmogony as a convenient pretext, Bayle has here skilfully suggested the dispensability of the metaphysical basis of Descartes's natural science, while stressing its radically naturalistic import.

[86] *ibid.*, n. "G," xi, 298.

Some time prior to Bayle's insinuations, Cudworth had undertaken a voluminous refutation of all varieties and shades of atheism both ancient and modern. In the latter category, Cartesians were appointed to a prominent place. After inveighing against Democritus and his Graeco-Roman successors, the English divine came to the more recent echoes of their doctrine in the *Principes de la philosophie*. Descartes's metaphysical safeguards had apparently been of little avail against the materialism spread by his physics:

"But it is prodigiously strange, that these Atheists should, in this their ignorance and sottishness, be justified by any professed Theists and Christians of later times, who . . . would feign persuade us . . . that this whole mundane system, together with plants and animals, was derived merely from the necessary and unguided motion of the small particles of matter, at first turned round in a vortex, or else jumbled altogether in a chaos . . . without the direction of any mind; God in the meantime standing by, only as an idle spectator of this *lusus atomorum*, this sportful dance of atoms—and of the various results thereof."[37]

By having replaced the Epicurean notion of fortuitousness in nature with that of a necessary regularity, definite advantages had, as Cudworth hastened to add, accrued to Cartesianism: "Nay, these mechanic Theists have here quite outstripped and outdone the atomic Atheists themselves." In view of the ineffectiveness of Descartes's ontological theology to counteract the lessons of his cosmogonical hypothesis, the *Intellectual System of the Universe* was forced to allude regretfully to "the Atheists . . . laughing in their sleeves and not a little triumphing, to see the cause of Theism thus betrayed by its professed friends and assertors."[38]

[37] Ralph Cudworth, *The True Intellectual System of the Universe: wherein all the Reason and Philosophy of Atheism is confuted*, New York, 1838, II, 93. The first edition appeared in 1678.

[38] *ibid.*, 94.

Among those who read Cudworth's remarks with care, John Ray, the noted English naturalist, likewise took to the field against materialism. By harnessing science to the task of inditing the *apologia* for Providence, he hoped to secure the menaced foundations of natural theology. Having first argued against the non-teleological outlook of classical Epicureanism, Ray accused "a sort of professed *Theists*, I mean Mons. *Des Cartes* and his followers" of perpetuating the ancient error "by excluding and banishing all consideration of final Causes . . . upon pretence, that they are all and every one in particular undiscoverable by us." His attention then focusing on the naturalistic implications of the *Monde*, Ray evaluated with disapproval the doctrine of Cartesians:

"They endeavour to evacuate and disannul our great Argument [i.e. the cosmological] by pretending to solve all the *Phaenomena* of Nature, and to give an Account of the Production and Efformation of the Universe, and all the corporeal Beings therein, both caelestial and terrestrial as well animate as inanimate, not excluding Animals themselves, by a sleight *Hypothesis* of Matter so and so divided and moved. [A synopsis of this follows]. . . . So that God had no more to do than to create the Matter, divide it into two parts, and put it into Motion according to some few Laws, and that would of itself produce the World and all Creatures therein."[39]

The heretical tendency of Cartesianism must, in the last quarter of the seventeenth century, have made considerable headway, for in 1692 the Jesuit Daniel's widely-read satire, *Le Voyage du Monde de Descartes*, took up at length several of the topics already discussed. Daniel was concerned in particular about the rivalry between natural science and

[39] John Ray, *The Wisdom of God manifested in the Works of the Creation*, 2nd ed., London, 1692, I, 24, 27-28. A French translation appeared in 1714, reprinted in 1729. Ray's contributions to botany and zoology have important places in the history of those sciences.

theology rising from the Cartesian background. The cosmogony outlined in the *Monde ou Traité de la lumière* appeared to conflict irrevocably with various articles of faith and with the order of divine providence. The Jesuit went to the root of the difficulty by saying that Cartesians proclaimed certain principles which were "extremely delicate and very dangerous with regard to questions related to our most holy mysteries. One does not know too well what to make of their opinions about the creation of our world, the production of matter, and the providence of God, who seems to have had no role other than that of making the little cubes of matter pirouette around their center. After that he had only to repose, since all the rest was able to come about without him."[40] Daniel discounted the general function of the theological basis of Descartes's physics, by means of which all events in nature had supposedly been related, in an ultimate sense, to God. Assuming, instead, that Cartesian natural philosophy began simply with matter in motion, he ridiculed the attempt to construct a world with such elements. The philosopher was quoted as saying: "I intend to execute here the system of my world . . . there you have matter, as much and more than I need; all that I need now is movement."[41] The eighteenth century, taking the cue mainly from Voltaire, was to summarize in almost identical words Descartes's physics; and concurrently, it was to characterize by the same formula the materialist science of Diderot and his colleagues.

Meanwhile, the Jesuit Daniel's satire reveals the goal that the main stream of naturalism, inspired by the *Monde's* example, was then pursuing: namely, the realization in the domain of physics of complete autonomy from theology. Daniel accordingly challenged an almost axiomatic proviso

40 Daniel, *Voyage du Monde de Descartes*, pp. 3-4.

41 *ibid.*, 47. Daniel gave a satirical paraphrase of Descartes's manufacture of a cosmos with matter and motion, pp. 295-306.

of Descartes's science: "Be it recalled what privilege Cartesians claimed when . . . they complained of being unfairly treated: to wit, that their philosophy made an abstraction of all questions of faith; that they were philosophers and not theologians; that they undertook to explain the mysteries of nature and not those of religion."[42] But in view of what was happening, the Jesuit was unwilling to grant Cartesians their desired separation of science from theology: "To say, as one does every day, that in philosophizing abstraction is made of religious subjects, and to reply thus only to all arguments demonstrating the opposition between certain principles essential to Cartesianism and the truth of certain mysteries of our faith—is this observing sufficiently the priority of theology; is it not, rather, a means of eluding it and refusing to submit to its yoke and authority?"[43] Père Daniel was convinced that Cartesians should not be allowed to have their way, for the problem involved, as he prophetically remarked, altogether "too dangerous consequences that favored heretics, infidels, and libertines." The attitude of the *Voyage du Monde de Descartes* about the subordination of natural science to revealed religion is important, for it is typical of the position the Church was to maintain during the whole course of the eighteenth century. By contesting the purely gratuitous jurisdiction claimed by their pious adversaries in a field where they lacked any real competence, the materialism of Diderot's group was to be the final, consistent outcome of the general policy and aim of Descartes's physical enterprise.

In addition to de Villars, Fontenelle, and Bayle, a certain *Traité des trois imposteurs*, which served as preface to a *Vie de Spinoza* by Dr. Lucas in 1719, may be cited as evi-

[42] *ibid.*, 40.

[43] *ibid.*, 526. In illustration of what he meant by this, Daniel said that the theologian should be consulted when it is obvious that "la définition Cartésienne de la matière n'est pas vraye, parce qu'elle ne peut s'accorder avec ce que la Religion nous enseigne touchant la création & la contingence de la matière." (p. 527)

dence of the alliance between Cartesian science and naturalism in the pre-Enlightenment period. The *Traité* emphasizes the inappropriateness of employing final causes in the elucidation of external phenomena, and vindicates the right of scientists to explore on their own terms the realm of things material, without patronage from theology.[44] The author contends that it is in reality the ignorance of their true causes that has ascribed physical events to supernatural agents: "It results therefrom that the person wishing to examine thoroughly the works of Creation, and to probe as a true scientist their natural causes without submitting to prejudices formed by ignorance, passes for being impious."[45] In the 1768 edition of the *Traité*, a commentary designed to clarify its meaning informs the reader that "this *Traité des trois imposteurs* is composed and reasoned out according to the method and principles of the new philosophy," which is identified with Descartes and Gassendi.[46] Also of interest is the unknown author's inclination, on the promptings of his Gassendo-Cartesian convictions, to fall into an expression, vague as it might seem, of pantheistic determinism.[47] The materialist ideology of the philosophes was likewise to reflect, in some measure, an accretion of Spinozist notions that served to consolidate and promulgate its basic tenets during their formative stage.

The naturalistic current stemming from Cartesianism which the Jesuit Daniel had attacked, then, continued to grow in the decades following. A specially significant instance of this may be seen in the notorious *Testament* of the free-

44 Anon., *Traité des trois imposteurs*, Yverdon, 1768, pp. 15-16. For the history and circulation of this work, see Ira O. Wade, *The Clandestine Organization and Diffusion of Philosophic Ideas in France from 1700 to 1750*, Princeton, 1938, pp. 124ff.

45 *Traité des trois imposteurs*, p. 18.

46 *ibid.*, 140.

47 *ibid.*, 22-23. A commentary (dated 1716) by a certain Pierre Frédéric Arpe states, p. 135: "l'Auteur explique à sa manière ce que c'est que Dieu, & en donne une idée assez conforme au systême des Panthéistes."

thinker Jean Meslier, which belongs most likely to the early years of the Regency, although its circulation in manuscript dates from the author's death in 1729. Meslier maintained that the conventional manner of interpreting Descartes was ill-founded in logic; that the Cartesian doctrine led instead, by way of pantheist conclusions drawn cogently from the definition of matter as extension, to an ultimate materialism and atheism.[48] As for Descartes's celestial physics, the *Testament* mirrored the unorthodox meanings taken to be inevitably bound up with it. A passage from Meslier suggests, too, that Malebranche's defense of the *tourbillon* hypothesis might well have abetted and diffused, against its intention, a germinating naturalism:

"According to the teaching of this author [Malebranche] . . . which is that of all Cartesians . . . it is clear and evident that the formation of the whole universe and the production of all the works of nature, and even their order, arrangement, and perfections could have resulted from natural forces alone, that is, by the motions of the particles of matter diversely figured, combined . . . [and] modified."[49]

Meslier adds that, according to the Cartesians, not only have the corpuscles of matter found their present arrangement by virtue of time and the laws of motion, but that if God were to put them in a different order, they would find their way back, as Descartes had declared, to the constitution which they now actually exhibit. In Meslier's comprehension of the *Monde* and *Principes*, the laws of motion would seem to confer certain self-determining powers to matter, in contrast to the essential passivity of *res extensa* asserted by Descartes's metaphysics. The *deus ex machina* is discarded from

48 *Le Testament de Jean Meslier*, éd. Rudolf Charles, Amsterdam, 1864, II, 367.

49 *ibid.*, III, 215-216. For an exact and succinct definition of the concept of Nature in Meslier, cf. Wade, *op. cit.*, p. 72. Professor Wade does not, however, trace to Descartes the historical source of Meslier's naturalism.

the consideration of the natural order which, characteristically, is assumed to proceed inexorably from the fixed mechanical laws, whatever they might be, of matter.

Another work, attesting the active role of Cartesian ideas in the spread of the materialist viewpoint, is the *Traité de l'infini créé*, which circulated in manuscript long before its publication in 1769. Almost certainly composed by the abbé Jean Terrasson who, with Fontenelle and Mairan, was reputed one of the eighteenth century's leading Cartesians, the *Traité de l'infini créé* purported, behind a mask of irony, to reconcile the most audacious implications of Cartesianism with the traditional dogma of religion. In reality, it served to disseminate materialistic and pantheistic thinking under a rather diaphanous apologetic disguise.[50] The printed version of the *Traité de l'infini créé* was put down slyly to the account of Malebranche, the editor capitalizing thereby both on the censor's approval for the unhindered distribution of the book, and on the pantheism latent in the Malebranchean system.[51] Terrasson's intention is to make plain the inevitable, logical conclusion urged by Cartesian philosophy, namely, that the Creation is infinite—a point that Descartes, it is said, had been aware of but had not openly affirmed.[52] Who

50 The fact that this work could have been taken by Bouillier (*Histoire de la philosophie cartésienne*, II, 610-616) as a sincere attempt to harmonize Cartesianism with Catholic dogma, is a rather glaring instance of how erroneously Descartes's influence on the growth of eighteenth-century naturalism has in the past been evaluated.

51 It was known to the philosophes that Malebranche, in trying to nullify certain unorthodoxies concealed in Cartesian thought, had himself become willy-nilly their spokesman rather than their critic. Maupertuis, for instance, wrote: "qu'on ouvre le livre de Descartes, qu'on voie de quelle manière il explique la formation du Monde, & ce qui s'en ensuivroit pour l'histoire de la Genese. Qu'on lise Malebranche, qu'on examine cette étendue intelligible archétype de tous les êtres, ou plutôt tous les êtres mêmes, que l'ame humaine n'apperçoit qu'en s'appliquant à cette étendue; qu'on suive cette idée, qu'on en tire les conséquences, & qu'on me dise ce que tout devient, ce que deviennent les corps, ce que devient la Bible." *Œuvres*, Lyon, 1756, II, 165*-166.

52 *Traité de l'infini créé* [par le Pere Malebranche de l'Oratoire], Amsterdam, 1769, pp. 1ff.

could be better suited, moreover, for stating this radical consequence than Malebranche who, according to an introductory "Vie du Pere Malebranche," was "more Cartesian than Descartes himself" and "was never frightened by the daring results to which that system could give rise!"[53] The *Traité de l'infini créé* first argues, as had Spinoza, that from the Cartesian concept of extension it follows that matter must be regarded as infinite, since it is not conceivable that extension as such could in any manner be assigned boundaries.[54] Likewise, matter is made infinite in duration, since the contrary of this would be equally unintelligible. To understand the eternity of material substance, a fundamental notion of Spinozism, the reader is told that "it is necessary to have accustomed his mind to the vast expanse that Descartes and his disciples have given to the universe."[55] The thesis central to the *Traité* is merged with that of Fontenelle's *Pluralité des mondes*: "We know very well that in admitting an infinite number of inhabited planets we are teaching nothing new to real Cartesians."[56] But while Fontenelle had had some scruples about describing the inhabitants of his pluralistic cosmos as human beings, owing to the difficulty of giving the Redemption an interplanetary scope, the author of the *Traité de l'infini créé* made no conciliatory gesture in this respect.[57]

A particularly revealing trait of Terrasson's book is that it tended, on the example of Cartesian cosmogony, to envisage the universe from its aspect of change as being in constant development by alternate destruction and creation. "For the vortices and their planets," reflected the *Traité*, "being abandoned to the general laws of motion, according to Descartes's explanation, change must necessarily take place regarding them, which will end in the destruction of each of these vortices." But the same causes will replace the

[53] *ibid.*, xxx. [54] *ibid.*, 8ff. [55] *ibid.*, 95.
[56] *ibid.*, 48. [57] *ibid.*, 50-51.

old with new worlds, "for the mechanical impact of elementary bodies, according to Descartes's theory which we suppose to be accepted, cannot but produce finally a sun and planets, since it is in this way that he assumes our world produced itself."[58] This conception of nature as something *in process* prepared the ground for the general theorizing of Diderot, Buffon, and their associates in the sphere of natural history. It furnished the ideological basis for such closely related speculations as Buffon's *Théorie de la terre*, and Diderot's transformistic view of organic origins. There will be occasion later to deal with this subject more amply, and to illustrate its ties with Cartesian precept. For the moment, it is sufficient to note that the *Traité de l'infini créé*, seeking to harmonize its developmental cosmos with the tradition of *Genesis* concerning Creation, effected the conciliation mostly at the latter's expense. Terrasson observed: "Some have explained the six days of which Scripture speaks as being six moments; but why should we not be permitted to interpret them as six years, six centuries, six thousand or six million years?"[59] It was presumably the same precedent set by Descartes's physics that suggested also to Montesquieu, in the *Lettres persanes*, the notion of a cosmos in continual becoming by virtue of matter in motion. Usbek summed up the orientation of the natural science he found in France towards 1720: "The world . . . is not incorruptible; even the heavens are not. Astronomers are eye-witnesses to their changes, which are the quite natural effects of the universal movement of matter. The Earth is subject, like other planets, to the laws of motion; it suffers within itself a perpetual combat among its forces; the sea and the continent seem to be at eternal war with one another; every instant produces new combinations."[60]

58 *ibid.*, 101-102.

59 *ibid.*, 118-119.

60 *Lettres persanes*, CXII, in Montesquieu, *Œuvres*, I, 355. Montesquieu was well acquainted with Descartes's philosophy of nature, to which he paid a glowing tribute in 1721: "Ce grand système de Descartes, qu'on

Turning to the province of clandestine manuscript literature, which nurtured the philosophes' most privately radical opinions, the themes already discussed are present, as might be expected, in a bolder and more articulate version. One such work, probably written around 1713, appears to belong in the direct line of thought linking scientific naturalism to Descartes's *Monde.* Its full title may be read in place of a table of contents: *Recherches curieuses de philosophie ou Dissertation sur les principes des choses naturelles, dans laquelle par le secours d'une méthode nouvelle on traitte de la génération des hommes, des animaux, des arbres, des plantes, de la formation du monde & de sa durée, des causes des vents, du tonnerre, de la foudre, de l'esprit, du raisonement, &t.*[61] A preface suggests that the "new method" in question might be the mechanistic conception in its universal applicability to the problem of nature.[62] While the need is asserted, from the first, of discarding or revising certain of Descartes's hypotheses, what actually follows remains, nonetheless, in essential agreement with the *Monde* by its resolve to clarify the formation of all things simply through the properties of

---

ne peut lire sans étonnement; ce système, qui vaut lui seul tout ce que les auteurs profanes ont jamais écrit; ce système, qui soulage si fort la Providence, qui la fait agir avec tant de simplicité et tant de grandeur; ce système immortel, qui sera admiré dans tous les âges et toutes les révolutions de la philosophie, est un ouvrage à la perfection duquel tous ceux qui raisonnent doivent s'intéresser avec une espèce de jalousie." VII, 48.

61 MS Bibliothèque Nationale, f. fr. 9107. This document, of 506 pages, purports to be a translation (1714) of the original by a certain "T.S.J.F." printed in London in 1713. While its previous publication is not certain, the manuscript's author appears to be a Dutchman and the dating reliable; its composition is clearly after 1700; cf., Wade, *op. cit.*, pp. 101-102. Professor Wade believes that the *Recherches curieuses* might well have entered France through the Boulainvilliers group, which was the rough equivalent, almost half a century earlier, of the "Côterie holbachique."

62 *Recherches curieuses*, p. 2: "au milieu du siècle passé Descartes se fit auteur d'une nouvelle philosophie; on ne peut assés le louër d'avoir osé mépriser les essences abstraites, les forces substantielles, les entités essentielles & une infinité d'autres expressions vaines qui ne signifient rien."

matter. The anonymous writer's cosmogonic project shows a faithfully Cartesian disinterest in teleological or supernatural factors. Its chief theoretical innovation—the making of "transformation of heat" the basic principle of physics—is of significance to us mainly by its conformity with the mechanical ideal.

But on several decisive issues, the *Recherches curieuses* leaves no doubt as to its immediate background in Cartesian science. Matter and extension are regarded as identical.[63] The *plenum*-theory implied by this[64] enters into the definition of Chaos, from which the world has evolved by a process of organic fermentation suggestive of Diderot's later speculations.[65] The universe whose origins are supposedly explained in the work turns out to be that of the vortices. Celestial bodies are pictured as supported by the fluid air surrounding them; their mode of motion is represented by the recurrent image of a piece of wood floating downstream; their distances from one another are regulated by the elastic resistance of the air compressed among them. This air is a *tourbillon*; the Sun and planets have each a *tourbillon* whose range of action results in the "capture" of lesser neighboring ones. The centrifugal force within a solar vortex, combined with the same force in reverse from outer vortices, serves to keep the various planets within their given system.[66] Inferences about a pluralistic cosmos ensue from these considerations; each fixed star is assumed to be a sun with inhabited planets of its own more or less like the earth.[67] In the Cartesian universe delineated by the *Recherches curieuses*, motion is conceived as "merely a consequence of the constitution of the body in which or by which it is pro-

63 *ibid.*, 250-251: "tous les espaces qui se présentent à notre esprit ne peuvent être que remplys de corps solides."

64 The interplanetary atmosphere is said to be a fluid mass "qui est la même dans tout l'univers & qui ne peut soufrir de vuide." *ibid.*, 251-252.

65 *ibid.*, 253ff.

66 *ibid.*, 292-314 *passim.*

67 *ibid.*, 353-354.

duced."[68] Such a universe, moreover, is held to be infinite, not in the positive, but in the negative sense,[69] which amounts to attributing with Descartes "indefiniteness" to its spatially inconceivable limits. In short, the philosophes could have found in the *Recherches curieuses* an interpretation of Cartesian physics that unequivocally anticipated their scientific naturalism.

Nearer in time to Diderot and his associates, another anonymous work in manuscript may be mentioned: the *Dissertation sur la formation du Monde*, dated 1738 and paralleling the *Recherches curieuses* in general theme and method.[70] The background of science seems, again, to be Cartesian. Matter is described as in essence extension.[71] The vacuum is denied.[72] The cosmos is depicted in relation to the *tourbillons* which, indeed, the *Dissertation* tends to accept in their main lines.[73] The specific theorizing of the treatise need not detain us, except to note that it emulates, in the origination of the world and its contents both inorganic and organic, the *Monde's* exclusive reliance on natural causes. A significant feature of the *Dissertation sur la formation du Monde*, moreover, is that it discloses, at a historically crucial moment, the intimate meaning of the naturalism to which, within its pages, Descartes's influence has led. After abstrusely discussing whether or not motion is essential to matter, the writer abandons at length this unsolved metaphysical puzzle as indifferent to his inquiry, and decides to regard matter simply as a substance invariably subject to change by reason of its possessing motion. In such a conclusion, Cartesian physics has lost completely its metaphysical preconditions: "I believe it therefore equally improbable that matter, considered in general, is absolutely deprived of motion and incapable of acquiring it, or that it has motion within itself as

68 *ibid.*, 249. 69 *ibid.*, 250. 70 MS Mazarine 1168.
71 *Dissertation sur la formation du Monde*, p. 28 and *passim*.
72 *ibid.*, 121. 73 *ibid.*, 121-122.

an essential property of its substantial nature. Perhaps we will come closer to the purpose by applying to this substance the idea we have of a thing which, with the aid of various preparations, becomes something else in appearance. It is from this third standpoint that we will treat our subject."[74] The author concedes to his critics that matter as a metaphysical abstraction does not imply movement; that movement need not, consequently, have inhered in matter from all eternity. He claims, nonetheless, that the actual concomitance of the two has the necessity purely of a constant physical fact, which in turn imposes the *scientific* necessity of adopting matter in motion as the first principle.[75] It was basically in the same sense that Diderot's naturalism, in the *Principes philosophiques sur la matière et le mouvement*, was to make of motion an "essential" property of matter.

Cartesian physics, as already indicated, was well-suited for fostering the philosophes' materialism by having converted to its own use certain fundamentals of Epicurean philosophy. This occurrence was distinctly familiar to the age of Enlightenment. The Jesuit Regnault seemed unwittingly to be justifying the materialist current that traced itself to Descartes when, in 1734, he drew in an objective spirit the analogy between Cartesian and Epicurean science. About the account of the world's origin in the *Monde*, he exclaimed: "Is not that the very language of Leucippus?"[76] Even the *tourbillon* theory was referred by Regnault to its sources in Leucippus and Epicurus, whose "innumerable worlds . . . were so many immense fluids, each of which revolved around its center and carried along its planets."[77] Among the philosophes, the marquis d'Argens testified to the underlying similarity, as seen by his contemporaries, between Epicureanism and Cartesianism: "These two doctrines, which seem very divergent

[74] *ibid.*, 54-55. [75] *ibid.*, 59-66.

[76] Noël Regnault, *l'Origine ancienne de la physique nouvelle*, Paris, 1734, II, 115-116; see also, II, 104ff. and *passim*.

[77] *ibid.*, 121.

because Gassendi admits the existence of a vacuum and Descartes denies it, have nevertheless a certain affinity with one another, by their agreement that the first principles of things go back to extremely fine corpuscles."[78] Although Descartes was not, strictly speaking, an atomist since he had maintained the indefinite divisibility of matter, this did not make much difference for D'Argens: "Descartes's subtle matter, except for its divisibility, is close enough to the atoms of Epicurus; and he uses it to just as good advantage as Gassendists do their hard and solid corpuscles."[79] Voltaire's *Dialogues d'Evhémère* expressed knowledge of the same kinship between Cartesian and Epicurean physics. Satirizing in the usual vein the construction of a world "with six-sided dice," Voltaire lamented that an entire nation was led to put stock in such extravagances, "just as Syracuse [believed] the absurdities of Epicurus: swerving atoms, animals formed from mud by chance, and a thousand other idiocies put forth with confidence."[80]

Diderot was no less aware of the parallel between Cartesian science and the speculations of classical materialism. Writing to Sophie Volland, he remarked about the *Rêve de D'Alembert* that he had considered naming its interlocutors Democritus, Hippocrates, and Leucippus; but that, not to confine the dialogue's meaning to the narrow limits of ancient philosophy, he had renounced the plan.[81] Now, Descartes's physics had succeeded in utilizing certain principles of Epicureanism within the modern context of a science built solidly

[78] Marquis d'Argens, *La Philosophie du Bon Sens*, nouv. éd., La Haye, 1768, I, 336.

[79] *ibid.*, II, 28.

[80] Voltaire, *Œuvres*, XXX, 502. Following the lead of Gassendi and Newton, Voltaire approved the atoms and vacuum of Epicurean physics; "mais pour le reste de la physique d'Epicure, elle ne paraît pas plus admissible que la matière cannelée de Descartes. C'est, ce me semble, se boucher les yeux et l'entendement que de prétendre qu'il n'y a aucun dessein dans la nature." *Dictionnaire philosophique*, I, 425.

[81] Letter dated 11 Sept. 1769; Diderot, XIX, 321.

on mechanics, and its authority had persisted well into the century of *lumières*. In conjunction with these facts, the link in Diderot's mind between Cartesianism and Epicureanism, as seen in the *Encyclopédie* article "Eléatique," assumes a special significance. Leucippus is there accredited with having originated the corpuscular conception, which Democritus and Epicurus later developed more fully: "He made everything proceed from the atom, its figure and motion."[82] And shortly after: "Although we have lost the works of Leucippus . . . sufficient knowledge has come down of the principles of his philosophy for us to judge the merit of some of our modern systematists; we could ask the Cartesians if the ideas of Leucippus are very far from those of Descartes."[83]

To characterize more exactly the naturalistic trend which, together with pantheist and atomist associations, had issued from Cartesian thought, one must juxtapose this with its most serious rival in the Enlightenment: the Newtonian natural theology. If, in the long interim from Descartes to Diderot, almost all scientists discarded final causes in their actual investigations, this did not commonly mean that once the desired facts were collected, and the appropriate hypotheses framed, they were not expected to square with certain theological preconceptions. Newton is a case in point. Voltaire thought that it would surprise many of his (presumably Cartesian-trained) readers that, "of all the proofs for the existence of God, the one based on final causes was the strongest in Newton's eyes."[84] The *Principia Mathematica* had truly succeeded in forging most intimate bonds between a deistic theology and mechanistic physics, with the result that Newton felt that a discourse on God "from the appearance of things does certainly belong to Natural Philosophy."[85] Père Castel, an anti-Newtonian proponent of vortex

[82] *ibid.*, xiv, 400. [83] *ibid.*, 402. [84] Voltaire, *Œuvres*, xxii, 404.

[85] Isaac Newton, *The Mathematical Principles of Natural Philosophy*, trans. into English by Andrew Motte, London, 1729, "Book Three, General Scholium," ii, 892. (The Latin first edition dates from 1687.)

physics in the eighteenth century, touched the crux of the matter when he said: "Descartes had the ambition of fabricating a world. Newton did not have the slightest desire in this regard."[86] Precisely that had kept the Englishman from recognizing any need to emancipate physics from theology. Newton's world, ostensibly the same in 1680 as it had always been since its contrivance by the Supreme Artificer, and doubtless what it would be until its dissolution with the millennium, presented no fundamental break with religious dogma. Accordingly the fact remains that, although the concern with natural theology had already a long history, Newton's teleological science, providing an ideal refuge for such ideas, inspired and substantiated a fresh wave of apologetic literature, both Christian and deistic, in Europe: Derham's *Physico-Theology*, Nieuwentyt's *Existence de Dieu*, the abbé Pluche's *Spectacle de la nature*, to mention only the foremost of an endless list.[87] All these works had in common the principal argument that patterns in nature attested special and discernible divine purposes.

The *Principia Mathematica*, viewing from a traditionalist standpoint the relationship of God to the order of nature, had expressed an "either-or" proposition concerning the status of finalism, which we might take as a curiously half-true prediction of what Cartesian science could be expected to bring about: "a God without dominion, providence, and final causes, is nothing but Fate and Nature."[88] Dr. Samuel Clarke, who expatiated on the theological possibilities of Newtonianism, was more explicit still. He claimed that Newton's ideas were irrevocably opposed to those of the material-

[86] Louis-Bertrand Castel, *Le Vrai système de Physique generale de M. Isaac Newton, exposé et analysé en parallèle avec celui de Descartes*, Paris, 1743, p. 18.

[87] This phase of Newton's influence has been discussed in a doctoral dissertation by Ruth T. Murdoch, *Newton's Law of Attraction and the French Enlightenment*, 1950. Microfilm copy in Columbia University Library, pp. 51-114.

[88] Newton, *op. cit.*, II, 391.

ists because, whereas the latter supposed the structure of the universe contingent upon the mechanical properties of matter, the former proved that its production was owing to an intelligent, free Cause.[89] Also Derham, in his *Physico-Theology*, was concerned about the well-being of established religion, "if we should allow the Atheist any of his nonsensical Schemes, the *Epicurean* his fortuitous Concourse of Atoms, or the *Cartesian* his created Matter put in Motion."[90] Although Derham charitably assumed Descartes's sincerity as an apologist for Christianity, the fact remained that his principles had by many been "perverted to Atheistical Purposes."

To understand in the rise of naturalism among the philosophes the crucial contribution of Cartesian science, it is necessary to state, regarding that event, the essential divergence that existed between Descartes's physics and the "orthodox" physics of Newton. The most effectual feature of the *tourbillon* system for the genesis of French materialism was, apparently, its having made the vortex a necessary consequence of a given definition of matter, rather than dependent, like Newton's law of attraction, on a supra-physical factor. From the identification of matter with extension, the theory of the *plenum* had logically followed; given the *plenum*, in which circular motion alone was possible, the *tourbillon* hypothesis was the ineluctable result; and so forth. Such a procedure, attributing all phenomena to the ultimate postulates of matter and its necessary motions, made Cartesianism the source of naturalistic tendencies which, it would seem, were sufficiently well-entrenched to be taken seriously by Newton's disciples. Roger Cotes, in a preface to the first English edi-

89 Dr. Samuel Clarke, *A Collection of Papers, which passed between the late Learned Mr. Leibnitz, and Dr. Clarke*, London, 1717, "2nd Reply of Clarke," p. 37. Readers in eighteenth-century France could easily have familiarized themselves with this material through Desmaizeaux's *Recueil de diverses pièces, par Mrs. Leibniz, Clarke, Newton*, Amsterdam, 1720.

90 William Derham, *Physico-Theology*, 12th ed., London, 1754, p. 271. This work, first published in 1713, was translated into French in 1732.

tion of the *Principia* in 1729, argued that vortex-theory, with its notion of the *plenum* and its ambition to explain natural effects on exclusively physical grounds, made no allowance for, and could not be fitted into, a teleological scheme of any kind. It was the persuasion of Cotes that in the end, the Cartesians would have to affirm that the constitution of the world was not "caused by the will of God, but by some necessity of its nature. Therefore they will at last sink into the mire of that infamous herd; who dream that all things are governed by Fate, and not by Providence; and that matter exists by the necessity of its nature always and every where, being infinite and eternal."[91] This prognosis, except for its insistence on the obsolete concept of Fate which, before long, will be replaced by the new idea of Nature, was not far from the mark. Cotes had, moreover, a rather clear perception of the particular conflict that was to typify the situation of natural science in France during the decades to come. He acknowledged that Newton's physics suggested something resembling occultism by its final reference to a cause outside nature, but countered that the only alternative to this, as exemplified by Descartes, was an atheistic science. About Cartesian critics of the Newtonian notion of physical causes, Cotes wrote: "They may call them, if they please, miracles or occult qualities; but names maliciously given ought not to be a disadvantage to the things themselves; unless they will say at last, that all philosophy ought to be founded in atheism."[92]

Voltaire was the apostle to the French of the Newtonian natural theology. Persevering throughout his long life as a *cause-finalier* (though on occasion recognizing its abuses), he was never to feel at ease with the unfinalistic orientation in science of La Mettrie, Diderot, Buffon, and their follow-

[91] Newton, *op. cit.*, "The Preface of Mr. Roger Cotes," unpaginated.
[92] *ibid.*

ers.[93] Voltaire's judgments about the naturalistic trend of the age are extremely valuable for the light they shed, indirectly, on the historical problem central to the present study. The satirical champion of deism decried most earnestly the method and aim of Descartes's physics, whose real import he summed up, derisively but correctly, to be: "Give me matter and motion, and I will make a world."[94] Elsewhere, Voltaire enlarged upon the meaning of his epigram through the medium of verse. Descartes was pictured as saying to God:

> "Quant à votre univers, il est fort imposant:
> Mais, quand il vous plaira, j'en ferai tout autant;
> Et je puis vous former, d'un morceau de matière,
> Eléments, animaux, tourbillons, et lumière,
> Lorsque du mouvement je saurai mieux les loix."[95]

Voltaire gave an estimate of the possible bearing of Descartes's enterprise, thus epitomized, on the emergence of naturalism in the eighteenth century, suggesting at the same time the mediating role of Spinoza's doctrine in the process by which Cartesian thought came gradually to fruition:

"The so-called principles of Descartes's physics are very far indeed from conducing the mind to knowledge of its Creator. . . . I say that the Cartesian system has produced that of Spinoza; I state furthermore that I have known

[93] The marginalia in his two personal copies of Diderot's *Pensées philosophiques* show that Voltaire's anti-materialism was based on the flat refusal to take matter and motion as the first principles of natural philosophy. Alongside "Pensée XXI" which gave the typical atheist's account of how the joining of corpuscles had shaped the cosmos, Voltaire wrote: "Vous supposez l'existence de ces dés—il est clair que rafle de six doit arriver; mais la question est, s'il y aura des dés: point de dés, point d'arrangement, point d'ordre. . . ." Cf., Norman L. Torrey, "Voltaire's Reaction to Diderot," *Publications of the Modern Language Association*, L (1935), 1128-1129, 1132.

[94] *Eléments de la philosophie de Newton*, 1738, in Voltaire, *Œuvres*, XXII, 404.

[95] *ibid.*, X, 169.

many persons whom Cartesianism has led to admit no God other than the immensity of things, and that on the contrary I have seen no Newtonian who was not a theist in the most rigorous sense."[96]

Voltaire's interpretation of the meaning of Descartes's science was not an isolated case. The lesson taught by the *Monde* was a pervasive one. The philosophes were commonly aware that, in this regard, Cartesian precedent pointed plainly towards materialism. Some years earlier, Montesquieu's Usbek had described the natural philosophy of the "Occidentals" in much the same style, but seemingly with an overtone of sly approval: ". . . left to themselves, and deprived of saintly marvels, they follow in silence the path of human reason. You would not believe how far this guide has brought them. They have disentangled chaos and explained by means of a simple mechanism the order established by divine architecture. The author of nature has given movement to matter: no more has been necessary to produce the prodigious variety of effects that we see in the universe."[97] Maupertuis recapitulated the Cartesian program in comparable terms, seconding however Voltaire's scorn: "Some philosophers have been foolhardy enough to attempt to explain, by means of these laws, its entire mechanism, and even its original formation: Give us, they have said, matter and motion, and we will form a world like the present one. A truly extravagant undertaking!"[98] La Mettrie, too, understood the radical antagonism between vortex-theory and theological tradition: "One fabricates vortices and cubes which explain everything, even that which is inexplicable—the Creation."[99] Although the *Abrégé des Systèmes* seemed skeptical about this Cartesian project, the materialism of the future

[96] *ibid.*, XXII, 404.
[97] Montesquieu, *Lettres persanes*, XCVII, in *Œuvres*, I, 310. Quoted in Bouillier, *Histoire de la philosophie cartésienne*, I, 194.
[98] Maupertius, *Œuvres*, I, 45-46. [99] La Mettrie, *Œuvres*, I, 238.

author of *l'Homme machine* and the *Système d'Epicure* was to find amplification and support in the general ideological outcome, among his contemporaries, of Descartes's cosmogony.

Like many at the time, Diderot had early in his career felt the force of the Newton-Voltairean theologizing of nature. The *Pensées philosophiques* of 1746, and to a lesser degree the *Promenade du sceptique* of the following year, had sought to refute pantheist thinking by virtue of a finalist interpretation of organic phenomena based on Newtonian example.[100] But the strong current of geological and biological interest during the 1740's, which emphasized the need to account for the formation of the world and its contents, rather than, in the manner of Newton, for its operations merely once Creation had taken place, inevitably drew Diderot, by progressive stages, from the orbit of Newtonianism into that of Cartesianism.[101] The *Lettre sur les aveugles* of 1749, in the few pages devoted to Saunderson's famous apocalyptic discourse, expressed the pith of Diderot's new system of nature. The cosmos was represented as being in a state of continual destruction and reconstruction by the laws of moving matter, and consistently with this, organic patterns were envisaged from a transformistic, developmental standpoint.

It was only normal that Voltaire should voice, as he ac-

100 Diderot, *Pensées philosophiques*, éd. R. Niklaus, Genève, 1950, pp. 12-13.

101 See my article, "From Deist to Atheist: Diderot's Philosophical Orientation, 1746-1749," *Diderot Studies*, ed. O. E. Fellows & N. L. Torrey, Syracuse University Press, 1949, pp. 46-63. The major shift in Diderot's approach to natural science evoked from the theologian Nicolas-Sylvestre Bergier a clairvoyant comment on the erstwhile deist of the *Pensées philosophiques*. "C'est à la connoissance de la Nature . . . qu'il était réservé de faire de vrais Déistes," he quoted Diderot as having held in 1746; but, "par une révolution singulière, cette connoissance ne fait plus que des Athées: voilà bien du progrès dans peu de temps." *Examen du matérialisme: ou Réfutation du Système de la nature*, Paris, 1771, I, 40. Depending whether nature was understood in the Cartesian or in the Newtonian sense, one was apt to be in Diderot's *milieu* either a materialist or a deist.

tually did in a polite note to Diderot, his lack of sympathy for Saunderson's views in natural philosophy.[102] Diderot's reply to this, defending the conception present in the *Lettre sur les aveugles*, formulated curiously enough the same kind of Spinozistic naturalism that Voltaire had previously attributed to the influence of Cartesian science. Diderot observed that, except for the threat of censure, he would have had Saunderson speak more openly. The starting-point of his argument would have been the Cartesian distinction of a spiritual and a material substance, neither of which, by Descartes's definitions, could be conceived as having any action on the other. Diderot's terminology, even, here becomes Cartesian: "Sound philosophy allows me to suppose in things only what I can perceive distinctly as belonging to them. But I perceive distinctly in spiritual substance no faculties other than those of willing and thinking, nor can I conceive how thought and will are able to act upon either material objects or non-being, any more than I can conceive of non-being and material objects having an effect on spiritual things."[103] The solution of this dilemma implied by Cartesian dualism falls in the line of Spinoza's pantheism: "To pretend that there can be no action by non-being and by material things on purely spiritual beings, because one is unable to conceive the possibility of such action, is the same as agreeing that there can be no action by purely spiritual beings on material objects, for the possibility of that action is similarly inconceivable. It ensues, therefore, from this admission and my reasoning, Saunderson would continue, that corporeal being is no less independent of spiritual being than the spiritual is of the corporeal; that together they compose the universe, and that the universe is God." Diderot's demonstration, it is true, does violence to good syllogistic form by its arbitrary conclusion. The author of the *Lettre sur les aveugles* has evidently argued here more effectively for the need to consider

102 Voltaire, *Œuvres*, XXXVII, 23. 103 Diderot, XIX, 421.

the sphere of nature independently of intelligent or final causes, than he has for a true pantheist doctrine. The concept of an autonomous Nature, to which Diderot's dialectic actually points, is the precondition of the type of natural science envisioned by Saunderson. The crowning bit of Spinozism is largely by way of blowing metaphysical dust in Voltaire's eyes. Diderot's remarks thus illustrate the peculiar current of naturalism, with its rather facile Spinozist accompaniments, to which Cartesian science had given birth.

The same filiation of ideas is attested also by D'Holbach's *Système de la nature* which, for motives of prudence probably, did not appear until 1770. D'Holbach insisted that from the general premises of Cartesian philosophy it followed necessarily that "there is no God other than nature, which is nothing but pure Spinozism."[104] On the basis of the specific principles posed by Descartes as foundations for his physics, D'Holbach held that "one is obliged to inform him that his system overthrows the notion of Creation." By contrast with this, he acknowledged that Newton's science was profoundly incompatible with pantheistic or naturalistic conclusions.[105]

Spinoza's thought, however useful for the propagation of the multiform ideology of eighteenth-century materialism, comprised neither the core nor the spirit of that doctrine. La Mettrie, Diderot, Buffon, D'Holbach were not primarily concerned with the elaboration of a naturalistic metaphysics. Central to the meaning of their naturalism was the vast plan of carrying to its completion the kind of unfinalistic, mechanistic science of nature of which Descartes, in the *Monde* and *Principes de la philosophie*, had sketched the broad outlines and defined the ultimate objectives.

104 Paul Heinrich Dietrich, baron d'Holbach, *Systême de la nature, ou des Loix du monde Physique & du monde moral*, Londres, 1771, II, 150. The "Bible of Atheism" also cited (pp. 151-152) in favor of its standpoint the overtones of pantheism which various critics had detected in the Malebranchean system.

105 *ibid.*, 153ff.

By 1739, the materialist current that was to culminate in the activities of certain philosophes appeared so plainly and unmistakably as issuing from the background of Descartes's natural philosophy, that the abbé Pluche was able to discuss the whole question openly in his *Histoire du Ciel*, and to call a spade a spade. Pluche's book offers complete corroboration, on all essential points, of the thesis that Cartesian physics led directly to the materialism of the Enlightenment. The abbé Pluche was well qualified to speak on the subject. His voluminous *Spectacle de la nature* was one of the most successful popularizations of scientific knowledge in the eighteenth century. As an ardent spokesman for Newtonian finalist theology, its author was particularly sensitive to the threat of naturalistic trends in science, and presumably well-trained for discerning their true character. The principal aim of the *Histoire du Ciel*, typical of an entire school of apologetics, to vindicate the "physics" of Moses may, perhaps, in itself be described as frivolous. Nevertheless, what the abbé had to say, apropos of his pious project, about the antagonistic influence of Cartesian science is extremely instructive. His analysis is sufficiently thorough and self-explanatory as to require no lengthy preface. Concerning the "Monde de Descartes":

"Not only is there no profit to be had from this imaginary physics, which pretends to relieve Providence of its role in the creation of the universe . . . but there is everything to be lost for man . . . [who] makes an idol of matter once it is set in motion. It is in reality blind, devoid of intelligence and design; nonetheless he attributes everything to it. It is matter in motion that engenders the elements. It is nature which has ordered the spheres, hardened the crust of the planets, and by a residue of lighter dust-like particles has surrounded the planet with an atmosphere. In a word, such a person, preoccupied always with this nature, hardly makes an occasional mention of the Prime Mover. He does not accede to Atheism, for that would be the height of folly. But the wisdom of the Deity, his intentions, foresight, goodness, and the constancy of his favors . . . find themselves absolutely banished from the greater scope of physics, and God is as much forgotten as if he had never existed."[106]

In Pluche's opinion, it is justifiable to use the general laws governing bodies in motion to explain the functioning of the already assembled cosmic machine and all its parts. But whenever the question is about the formation of things, reference must be made, not to nature as such, but to as many special designs as there are pieces in the phenomenal order. Within limits, it is true, the immediate causes for specific effects may be looked for and given in terms of mechanism; but the primary causes for the production of the elements, such as water, fire, iron, etc., are not to be elucidated except in relation to the express wishes of the Creator.[107] Descartes's physics, running counter to this traditional scheme by its effort to manufacture a world complete with furnishings from matter and motion only, served, in Pluche's judgment, as an incentive to atheism (a word sometimes made loosely synonymous with naturalism). The abbé observed about his contemporaries: "This method of employing only movement for the organizing of matter is what has inspired the greatest confidence in atheists, by leading them to accept as profound physics a few appearances of rational thought accompanied by geometry."[108] A friend of the abbé's, moreover, having frequented the atheistic circles which the

[106] Abbé Noël-Antoine Pluche, *Histoire du Ciel, considéré selon les idées des poëtes, des philosophes, et de Moïse*, Paris, 1739, II, 255-256. Although it is not known when Diderot first became acquainted with this book, his *Encyclopédie* articles "Divination" and "Polythéisme" indicate that he read it carefully. Diderot, XVI, 359ff.; XIV, 290-291.

[107] Pluche, *op. cit.*, II, 257-260. Descartes's basically non-finalistic cosmogony was made the equivalent of "saying with Lucretius that the eye has not been made to see." *ibid.*, 245.

[108] *ibid.*, 262.

Histoire du Ciel indicts, has provided the following information about the sort of argument to be heard there:

"There is, they say, an absolutely indifferent universal matter which is susceptible to all kinds of forms. . . . Let us be free to suppose, for a moment, that this matter is eternal and to add to it a movement which will be eternally distributed through all its parts. This suffices us for the explanation of all things; and if this is capable of sufficing, why should we wish to admit any other principle? For deciding to refer conveniently the world's organization to an eternal motion rather than to an eternal wisdom, it is enough that this organization could be the simple effect of movement. Now, the thing is evidently possible. The great Descartes, the most reflective and systematic mind, and the most accustomed to admit only what can be conceived with evidence, took as the basis and first principle of his whole physics, that matter *in motion ought to produce all the objects, both general and particular, which are to be seen in the world, without God giving it any order or proportion.* These are his very words."[109]

Interestingly, Diderot's *Pensées philosophiques* were to ascribe a similar line of reasoning, reinforced by a theory of mathematical probability, to the standard type of atheist against whom that work, still faithful to Newtonian deism, was to be in part directed.[110] As for the abbé Pluche, the conclusion drawn by him from the observed cooperation between Cartesian science and atheistic materialism is perfectly lucid:

"What appears to me most clear and certain about this reasoning of the atheists is the shame that falls, as a result of it, on that ostentatious system of physics, which asks for only matter and motion in order to construct the world.

[109] *ibid.*, 263-264.
[110] Diderot, *Pensées philosophiques*, pp. 16-18.

Materialism is the outcome of it. But those who hope to authorize atheism and all its deplorable consequences by means of Cartesian doctrine have had recourse to dreams and fictions for the purpose of upholding impiety. Their demand for a material substance which, from all eternity, gives movement to itself, is an absurd request. . . . Furthermore, they imagine that they are able to put into play the homogeneous and universal matter of Descartes, because by the wearing away of its [corpuscular] surface, and by its spinning about in a vortex, it becomes anything they want it to become."[111]

As opposed to the close bond between Descartes's *Monde* and the growth of materialist ideology during Diderot's youth, Pluche paid tribute to the radically different import of Newtonian physics. This latter was declared to be "perfectly in accord with the account of Moses, inasmuch as Newton, like Moses, related the engendering of the various elements and the arrangement of the whole to as many wishes and dictates of the Creator, and not to any physical cause."[112] But there was already a tendency, in France, to employ even the laws of attraction within the broad theoretical framework of Descartes's cosmogony, and to try to do with attraction what the *Monde* had not succeeded too well in doing with the *tourbillons*. This temptation was to grow stronger with the years, as best seen perhaps in Buffon's *Théorie de la terre*, after attractionism had consigned the vortex hypothesis more or less to oblivion. Pluche was far-sighted enough to class all such efforts, from the start, for what they were in reality to be: "To say that attraction has formed the Earth and planets is to fall back into those imaginary constructs such as we have from Lucretius and Descartes."[113] In this respect, an essential divergence ex-

111 Pluche, *op. cit.*, II, 265-266. 112 *ibid.*, 274.

113 *ibid.*, 318. Buffon and Maupertuis were, of course, to make just such a use of attractionism, although the latter's interest centered in biological phenomena more particularly, and left, as Newton had done, the problem

isted between Cartesian thought and Newtonianism properly understood. In the light of this divergence, the authentic sources of French naturalism in the Enlightenment may be distinguished from its various surface accretions and secondary features.

In Descartes's physics, nature and its laws derived, so far as scientific inquiry was concerned, ultimately from the definition of material substance; the constitution and functioning of the cosmos were considered inherent in the properties of moving matter. The dependence of the universe on God, however certain, was not a subject for investigation by science; about the ineffability of the cosmological link, Descartes had written to Mersenne: "I know that God is the author of all things. . . . I say that I know it, and not that I either conceive it or understand it."[114] Such a conception of nature led, both logically and historically, to the philosophes' scientific naturalism. But there was prevalent during the eighteenth century another envisagement of nature, which

of cosmic origins unresolved. A good example of the application, within the framework of Cartesian cosmogonical science, of Newton's law of gravitation was Pierre Estève's *Origine de l'univers expliquée par un principe de la matière*, Berlin, 1748. Kantian philosophy would seem to have had its historical roots partly in the same general trend. When the *Kosmogonie* (1755) simplified Descartes's formula to "Gebet mir Materie, ich will eine Welt daraus bauen," Kant was expressing the concomitance of motion with matter that had already resulted from the evolution of Cartesian physics. Pluche's shrewd prediction of the probable synthesis between Newtonian astronomy and the broader objectives of the *Monde* was to find, towards the end of the century, its greatest exponent in Laplace. Although he contributed signally to perfecting Newton's world-system, one detects the purely Cartesian inspiration behind the goal proposed by Laplace for astronomical inquiry: "Cet arrangement des planètes ne peut-il pas être lui-même un effet des lois du mouvement, et la suprême intelligence, que Newton fait intervenir ne peut-elle l'avoir fait dépendre d'un phénomène plus général? Tel est, suivant nos conjectures, celui d'une matière éparse en amas divers dans l'immensité des cieux. . . . Parcourons l'histoire des progrès de l'esprit humain et de ses erreurs, nous y verrons les causes finales reculées constamment aux bornes de ses connaissances." Quoted in Bréhier, *Histoire de la philosophie*, Tome II, fasc. 2, p. 315.

114 Letter dated 27 May 1630, in Descartes, I, 152.

was not always adequately distinguished from the Cartesian. This was the view that assumed, as evidenced by Newton, Leibniz, Voltaire, Rousseau, Bonnet, and others, that the assemblage and working of the cosmos were to be referred, in a final sense, to an outside, intelligent agent. This Newton-Voltairean approach endeavored to fit the *ensemble* of physical laws governing the universe into a teleological scheme. The world was a machine that had been contrived by the Supreme Craftsman, and its operation had the passivity of a clock's.

The Cartesian "world" was a machine too, but a machine that had *developed* from matter in motion and could be expected to keep changing indefinitely with the subsistence of its primary causes. Descartes had plainly declared, moreover: "I consider that there are an infinity of diverse movements existing perpetually in the world."[115] This Cartesian formula for the conservation of momentum in an equal and constant quantity (*mv*) was destined, despite its being erroneous, to prompt and support the naturalistic outlook. It was Maupertuis, the first Newtonian physicist in France (amicably nicknamed by Voltaire "*sir Isaac* Maupertuis"),[116] who testified to the fact. He classed with the Cartesian principle of thermodynamics Leibniz's modification of it into the notion of a *vis viva* which, more correctly preserving a constant quantity in the collision of bodies by virtue of multiplying their mass into the square of their velocity, seemingly undermined the finalism that the German philosopher sought elsewhere to establish.[117] Newton, on the other hand, showed

115 Descartes, XI, 10.

116 Voltaire, *Œuvres*, XXXV, 54.

117 Maupertuis, *Œuvres*, II, 240. For an excellent treatment of the main issues involved in the Cartesian-Leibnizian controversy over the view of "constancy" in the universe, consult H. W. B. Joseph, *Lectures on the Philosophy of Leibniz*, Oxford, 1949, pp. 27-61. The divergence between the concepts of "motion" and of "force," as implied respectively by the different formulae of Descartes and of Leibniz, does not alter the focal question before us, i.e., the derivation of the philosophes' materialism

that there was more often loss than conservation of energy in mechanical process, and "believed that in the end it would completely die out, if God did not from time to time communicate new forces to the machine of the world. This idea appeared hardly philosophical to those who wished to exclude the world from the rule of Divinity."[118] Maupertuis's own "law of minimum action" was intended as an antidote for the naturalistic tendency of Descartes's science, in which Leibnizian thought had shared to some extent. "Either of the so-called principles of Descartes and Leibniz, *the conservation of the quantity of movement* or *the conservation of the active force*, would attribute eternity and independence to the motions of the universe"; whereas the law of minimum action, in the tradition of Newtonian teleology, "makes it apparent that these movements are neither eternal nor independent; that they are subject to a power that produces and augments, then diminishes and destroys them in the wisest and most economical fashion."[119] In terms of Maupertuis's analysis, Diderot may be said to have expressed, in the *Principes philosophiques sur la matière et le mouvement*, a viewpoint combining the Cartesian and Leibnizian theories of constant momentum and constant force in the universe.[120] Compared

from Cartesian physics. Leibniz was an important intermediary in that transmission of ideas, even if in other respects his philosophy of nature endeavored to forestall the naturalistic conclusions suggested by Descartes. But so far as materialist ideology was concerned, the opposition between Cartesians and Leibnizians amounted for the most part, as Maupertuis realized, to a logomachy.

118 Maupertuis, *Œuvres*, II, 240.

119 *ibid.*, 242.

120 Diderot, II, 66-67. "La quantité de force est constante dans la nature; mais la somme des *nisus* et la somme des translations sont variables. Plus la somme des *nisus* est grande, plus la somme des translations est petite; et, réciproquement, plus la somme des translations est grande, plus la somme des *nisus* est petite." Diderot would seem here to be working towards a theory of interdependence and correlation between the notions of "translation" (identifiable with the Cartesian definition of motion) and of "nisus" (corresponding to the Leibnizian idea of active force). The same viewpoint was woven into D'Holbach's *Système de la nature*, I, 19-21.

with the "self-winding" cosmos of Descartes, the eighteenth century was to see nothing more than an ordinary winding-clock in the Newtonian world. It resulted that Nature, while for Newton's followers something essentially created, passive, and static, came to be understood in the complex of Cartesianism as something creative, active, and dynamic. It was the latter conception that underlay the scientific naturalism of the philosophes by consistently emphasizing the problem of development in the physical universe, and by offering a key to its investigation. The interests in science of the materialist school were to center, accordingly, in cosmogony, geology, and biology.

A concrete instance of how Cartesianism prompted the idea of a creative Nature may be seen in the abbé Terrasson, already mentioned in connection with his pantheistic *Traité de l'infini créé*. D'Alembert's academic eulogy of him acknowledged that, although not zealous about any one system of philosophy, "Cartesianism was that which he seemed to have adopted."[121] Certain reflections by Terrasson, contained in a posthumous publication of 1754, belong in all likelihood to that earlier phase of eighteenth-century naturalism to which the abbé Pluche called attention. Terrasson's Cartesian leanings were apparently not of the official kind. To both Newtonians and orthodox-minded Cartesians, neither of whom wished to assign a purely physical first cause to either gravitation or the vortices, Terrasson proposed the notion of a "mouvement Inné."[122] The theory of an inherent motion in Nature he contrasted with the altogether too clocklike world of Newton and the teleology with which it was bound up: "There is in Nature a perpetual movement differing from that which is sought with regard to machines. The latter could be procured only from the outside, whereas the former

121 D'Alembert, *Œuvres*, III, 373.

122 Jean Terrasson, *La Philosophie applicable à tous les objets de l'esprit et de la raison*, Paris, 1754, p. 203.

is innate. This innate motion maintains the course of the stars that are actually formed; and sustains that fermentation which is the cause and principle of animal life and of the generation of all the living things that exist or will exist to the end of time."[123] The universe that had taken shape from the Cartesian postulates of matter and motion was a universe of continual process, rather than of fixed and static patterns. In the representation of it, development rather than structure was stressed: "Universal matter, accompanied always by movement, having formed until now suns and planets, will go on forming them, similarly, to the end of time. But these latter will no longer be the same suns nor the same planets as the first, for none of these things are able to last forever."[124] Such attitudes as Terrasson's were not far removed from the core of Diderot's naturalism as initially expressed, for example, in the Saunderson-episode of the *Lettre sur les aveugles*.

Having traced to its main source the materialism prevalent in France up to about 1750, a recapitulation of its broadest features discloses (though at the risk of oversimplification) the elements of Cartesian thought brought to the philosophes' fore-consciousness by the late 1740's. The first of these was the resolve to expound the origins of the entire order of nature, from its cosmic outlines down to its minutest organic details, in terms exclusively of matter in motion. Such a project involved, in second place, the complete rejection of final causes, or providentialism both general and particular, in the interpretation of physical phenomena. Thirdly, it ensued from these two aims that the total separation of natural philosophy from theological prerequisites became a *sine qua non* for guaranteeing the future progress of scientific inquiry.

How had Descartes himself dealt with these crucial issues? The *Monde ou Traité de la lumière* was meant, by its author's own declaration, to reconstruct, from matter and motion

123 *ibid.*, 204-205.

124 *ibid.*, 212, and ff.

only, a hypothetical model of the actual world. Cartesian physics rested on the single-minded use of efficient cause defined in the sense of mechanical impact or *choc*. In itself, however, the reliance on efficient cause was quite common to the birth of mechanistic science in Descartes's period; on that ground alone, Gassendi might likewise be taken as the fountainhead of the philosophes' naturalistic science. At first glance, Gassendi would seem indeed to be Descartes's most serious rival in that regard. But closer scrutiny reveals the latter's potent originality. Gassendism, although it corroborated and promoted the influence of Cartesian ideas in the eighteenth-century *milieu*, could not in effect have inspired, in as logically conclusive and coherent a sense as Descartes's science did, the materialism of Diderot's circle. Bernier's *Abrégé de la philosophie de Gassendi*, through which the major share of Gassendism filtered down to the Enlightenment (for the *Syntagma* itself hardly invited perusal), followed its model in treating the problem of nature and finalism in a confusing and ambivalent fashion. On the one hand, Gassendi had proposed to explore nature in terms of efficient cause; on the other, his natural philosophy had given a prominent place to final causes, divine providence, and the proof of God's existence from the "contemplation of nature."[125] On these decisive questions, the *Abrégé* reflected Gassendi's break with Epicurean tradition and his insistence on the teleological factor in physics: ". . . how could the Atoms have taken hold, hooked together and united themselves firmly enough in order not to dissolve again and escape forthwith into the immensity of space, returning thereby to their original state of chaos; and how finally could they have disposed themselves so happily as to form this marvelous machine of the world, if some very powerful and intelligent

[125] François Bernier, *Abrégé de la philosophie de Gassendi*, Lyon, 1684, II, 209-231, 242-257. Although the *Abrégé* is often an unreliable guide to Gassendi, on the points here under discussion, needless to say, no significant deviation from the original took place.

Cause had not joined, united, and disposed them in that manner?"[126] Under the circumstances, French materialism could not have derived primarily from Gassendi without incurring confusion about its basic meaning. By contrast with his contemporary, however, for the author of the *Méditations métaphysiques* the utter incomprehensibility of God's intentions vitiated from the start any reasoning about nature through finalism:

". . . knowing already that my own nature is extremely weak and limited, and God's on the contrary is immense, unfathomable, and infinite, I have no difficulty in realizing that there are an infinity of things in his power whose causes surpass the bounds of my understanding. And this reason alone is sufficient to persuade me that all that type of cause customarily based on ends is of no use at all in physical and natural matters; for it does not seem that I can without temerity inquire into and attempt to discover the impenetrable purposes of God."[127]

Clear as the meaning of this Cartesian manifesto was, almost a century passed before its full implications were brought to fruition. It was towards 1750 that the irreducible conflict originating in Descartes's physics between a finalistic and a non-finalistic natural science came to an inevitable climax. Maupertuis's *Essai de Cosmologie* described this situation quite explicitly at the precise historical moment when scientific naturalism was in its birth-throes: "All philosophers and scientists today fall into two sects. One group would like to subject nature to a purely material order, to exclude from it all intelligent cause, or at least wishes that, in the explanation of phenomena, reference never be made to such a principle, and that *final causes* be entirely banished. The other, on the contrary, makes continual use of these

[126] *ibid.*, 250.

[127] *Méditations métaphysiques*, iv; Descartes, IX, 44. Also, *Principes de la philosophie*, I, 28; IX, 37.

causes, discovers throughout the whole of nature the Creator's purposes, penetrates into his designs in the minutest phenomena. According to the former, the universe could get along without God. . . . According to the latter, its smallest parts demonstrate his agency. . . ."[128] Maupertuis's personal misfortune was to be caught on the horns of the dilemma whose character he so distinctly perceived; for his own thinking contributed somewhat ambiguously, as will later appear, partly to the first group, and partly to the second.[129] Be this as it may, the unfinalistic orientation stemming from Cartesian science culminated, eventually, in the naturalism of Diderot, La Mettrie, Buffon, and their colleagues. The integral sense of Descartes's above-quoted lesson is present in the *Lettre sur les aveugles*, where it occurs appropriately in conjunction with the earliest, embryonic statement of Diderot's system of nature:

"Is a phenomenon, in our estimation, above human grasp? We say immediately: *it is the work of a God.* Our vanity will not be satisfied with less. Could we not put in our pronouncements a little less arrogance, and a bit more philosophy? If nature offers us a knot that is difficult to untie, let us accept it for what it is. Let us not employ in cutting it the hand of a Being who then becomes for us a new difficulty even more insoluble than the first."[130]

In the task of liberating natural science from the tutelage of theology, which the materialists carried out uniformly with the general rejection of finalism, Descartes had pointed

128 Maupertuis, *Œuvres*, I, iv.

129 Maupertuis described the reactions of the two opposing schools to his own efforts to find a middle course: "Comme il n'y a aujourd-hui presqu'aucun Philosophe qui ne donne dans l'une ou dans l'autre de ces deux manieres de raisonner, je ne pouvois guere manquer de déplaire aux uns & aux autres. . . . J'ai été attaqué par ces deux especes de Philosophes, par ceux qui ont trouvé que je faisois trop valoir les causes finales, & par ceux qui ont cru que je n'en faisois pas assez de cas." *ibid.*, v-vi.

130 Diderot, *Lettre sur les aveugles*, éd. R. Niklaus, Genève, 1951, p. 40.

out the way. The author of the *Monde*, perhaps sensing the conflict between science and religion that might well spring from his program in physics, had indicated basically the same solution that Diderot and certain philosophes were to reiterate. Descartes, doubtless taking courage from the protective virtues of his metaphysics, had been bold enough to deny any scientific value to purely theological materials—for instance, Scripture: "I believe . . . it is an application of Holy Scripture to an end for which God did not intend it, and consequently an abuse of it, to wish to draw therefrom truths which belong to the domain of the human sciences only."[181] An abuse of the type that Descartes had in mind was, specifically, the substitution of the account of the world's creation in *Genesis* for a scientific explanation in physical terms. The *Principes de la philosophie*, dealing with this extremely delicate question (of the kind that had already postponed indefinitely publication of the *Monde*), made the dichotomy between the spheres of theology and physics as clear-cut as ecclesiastical authority in the seventeenth century would have tolerated. Descartes's command of ironic subtlety for the occasion was of a quality that the philosophes themselves might well have admired. About the cosmogony in *Genesis*, Descartes started by saying: "The Christian religion wishes us to believe it to be so, and natural reason persuades us absolutely of this truth." Yet he was far from accepting the Mosaic physics as the last word on the subject; as prudently as possible, Descartes justified the ambition of the scientist to give his own explanation of how things had come to be what they are: ". . . just the same, we could render more comprehensible what is generally the nature of all

181 Descartes, II, 347-348. Descartes's attitude was not an isolated case. For example, Mersenne among his contemporaries, with much placidity (and some ingenuousness), combined a complete trust in mechanistic philosophy with a sincere reverence of Scriptural and other dogmas. Cf., Robert Lenoble, *Mersenne, ou la naissance du mécanisme*, Paris, 1943, pp. 223ff.

things that are in the world, if we were to imagine several very intelligible and extremely simple principles, by means of which we were to show clearly that the stars and Earth and, all in all, the whole visible world could have been produced as if from several seeds, although we know that it was not produced in this manner."[132]

A century later, Descartes's proposal, taken with its equivocal reservation, became a commonplace device of materialist practice. About its recurrence as a preface to both the *Théorie de la terre* and the *Epoques de la nature*, for instance, La Harpe was to remark, in 1797, that atheists were not less inclined for that reason to claim Buffon as one of themselves.[133] With tongue in cheek, Diderot offered his theory of transformism with an identical accommodation: "Religion spares us many errors and much labor. If it had not enlightened us concerning the origin of the world and the universal system of beings, how many different hypotheses would we have been tempted to accept as the secret of nature!"[134] Those critics who have minimized the anti-traditionalism of the *Monde* and *Principes* by asserting that Descartes's cosmic system was offered as mere hypothesis and not as fact, have overlooked that the hypothesis was intended as an *interpretation of fact.* As such, Descartes's cosmogonic method thrust a dilemma upon the natural science of the future. If the Biblical origin of the world was factually true, the *Monde* and all cognate speculations were at once futile in their objectives and misleading in their conclusions; but if the physicist was free to explain the world's formation by exclusively mechanical causes, theology would be obliged eventually to renounce its presumed competence in that domain. During the Enlightenment, the naturalistic ideology that had its egress from Cartesian thought resolved the

[132] Descartes, *Principes de la philosophie*, iii, 45; IX, 123-124.

[133] Jean François La Harpe, *Cours de littérature*, Paris, 1825, T. XVII, "De la philosophie du XVIIIe siècle," p. 69.

[134] Diderot, II, 58.

dilemma in the expected manner. The marquis d'Argens commented on the author of the *Monde*: "It would have been desirable that this philosopher had lived in the time of Moses; he would have given him some excellent advice. For the Hebrew prophet knew nothing about the spinning around of square-shaped particles, or at least he does not mention a word about it in *Genesis*."[135] It is one of the curious ironies of history that Descartes had hesitated to publish the *Monde* in his lifetime, not because of the thoroughgoing naturalism that some readers were to find in its pages, but because it contained the then heretical, but otherwise innocuous, belief in the earth's motion.

To realize how "advanced" Descartes's position had been for the seventeenth century, one might consider how Gassendi had treated the same problem. Bernier's *Abrégé* formally repudiated those aspects of Epicureanism bearing upon the world's self-production from atoms in motion. The reason given was that "the light of Holy Scripture serves us here as guide, and teaches us the manner in which God in the beginning created, separated, and embellished all things; so that it does not seem that one ought to ask for anything additional."[136] If Gassendi had replaced by the agency of God the *Fatum* of classical tradition, this did not render the process of Creation any more intelligibly determinable, i.e. the object of mechanistic science, than it had been in the past. In Bernier's presentation of Gassendi's ideas, the world's formation in the physical sense was hardly distinguished from its creation in the theological sense; and an enterprise such as Descartes's cosmogony would have been considered beyond the reach of human intelligence.[137] It is not likely that, in

135 D'Argens, *Philosophie du Bon Sens*, II, 27.

136 Bernier, *Abrégé*, III, 381.

137 *ibid.*, 380-386. An awareness of the contrast between Gassendi and Descartes on these points is to be seen, for example, in René Rapin's *Reflexions sur la Physique*, which described the first thinker as: "du reste Epicurien mitigé par principe de conscience. Car il avouë la creation des atomes qu'Epicure nie, il veut que Dieu leur donne le mouve-

their efforts to free natural science definitively from theological domination, Gassendism could have flattered greatly the hopes of eighteenth-century materialists.

By contrast with all this, Descartes's cosmogonic speculation was bound, sooner or later, to bring into conscious opposition the divergent approaches and aims of science and theology. The intensification of geological and biological interests in the 1740's, by centering attention on the developmental aspect of phenomena, proved a potent factor in bringing about the eventual triumph of the naturalistic implications of Cartesianism. Envisaging physical fact from a Heraclitean perspective of change, and substituting a cosmocentric science for an anthropocentric one, Descartes had already underlined the inappropriateness of any specifically theological or providential scheme of Creation:

". . . it is in no way likely that all things have been made for us, and that God could have had no other purpose in creating them. It would be impertinent, it seems to me, to wish to make use of such an opinion in support of reasonings about Physics. For we could not doubt that there are an infinity of things at present in the world, or which have once been there and have since entirely ceased to exist, without any human being having ever seen or known them, and without their ever having served him to any purpose."[188]

Descartes's viewpoint, translated into the ideology of certain philosophes, resulted in the definitive demarcation of the different spheres proper to science and to theology. It is not far from the words of Descartes to Diderot's exclu-

ment, l'extension, & la figure, qu'ils ont d'eux-mesmes dans Epicure: il admet la providence, que ce Philosophe ne reconnoist pas." Père Rapin's estimate of Descartes—"tout Sectateur de Democrite qu'il pretend estre"—is not as indulgent: "quand, pour rendre raison des choses, il a dit qu'elles se font par une certaine figure, par un certain mouvement, par une certaine extension, il a tout dit." *Œuvres*, Amsterdam, 1709, II, 422, 425.

188 Descartes, *Principes de la philosophie*, iii, 8; IX, 104.

sion of all teleological considerations from the scientific program of the *Interprétation de la nature*:

"Who are we to explain the ends of nature? Do we not perceive that it is almost always at the expense of her powers that we extol her wisdom; that we reduce her resources by more than what we could ever make up to her intentions? This manner of interpreting nature is bad, even in natural theology. It amounts to substituting man's conjecture for the work of God. . . . The physicist . . . will therefore abandon asking *why*, and will busy himself only with the question: *how*."[139]

In the two passages juxtaposed, an apparent dissimilarity will have struck the reader: where Descartes spoke of God, Diderot alludes to nature. Descartes, in effect, had formally averted the accusation of materialism by conceiving of matter as in itself inert, and requiring the agency of God to set it in motion. In the *Principes philosophiques sur la matière et le mouvement*, Diderot challenged this conception, and chose instead to include motion and energy in the essence of material substance.[140] This divergence, profound as it may seem to the student of metaphysics, proved actually irrelevant to the influence exercised by Cartesian physics. Once God had communicated motion to matter, the *Monde* had sought to clarify the emergence of things without further reference to the Prime Mover. Unlike the role Newton reserved to the Supreme Artificer, the assembling of Descartes's world-machine was the responsibility of "la nature

[139] Diderot, II, 53-54. In the article "Du Cochon" of the *Histoire naturelle*, Buffon blamed in similar terms the finalistic bias in natural science, from which it results that "nous imaginons des rapports qui n'ont aucun fondement, qui n'existent point dans la nature des choses, et qui ne servent qu'à l'obscurcir: nous ne faisons pas attention que nous altérons la philosophie, que nous en dénaturons l'objet, qui est de connaître le *comment* des choses, la manière dont la nature agit; et que nous substituons à cet objet réel une idée vaine, en cherchant à deviner le *pourquoi* des faits, la fin qu'elle se propose en agissant."

[140] Diderot, II, 64ff.

seule."[141] The relationship of God to nature so defined, by transcending the limits of physical science, was of no more interest to the philosophes than other metaphysical or theological problems. Diderot, like Descartes, would have been willing to describe the mechanical order in the most general sense as inscrutably the "work of God," provided however that only nature itself, free of any providential plan, was made the proper object of intelligible knowledge. In the science both of Descartes and of Diderot's group, there intervened between God and the realm of phenomena the notion of a demiurgal Nature, having creative, self-determining powers. Descartes had defined nature as "matter itself"; metaphysically, the essence of *res extensa* was conserved by God, but in the specific domain of physics the various motions and combinations of its particles were attributed instead to Nature, since God's immutability could not be reconciled logically with the perpetual flux of the material universe.[142] For Diderot, nature was "the actual general result or the successive general results of the combination of the elements of matter."[143] Both Descartes and Diderot, in the theoretical reconstruction of the external world, relied on the laws and properties of a creative Nature or, simply, of matter in motion.

To understand more concretely how materialistic Cartesianism culminated in the ideology of Diderot and his associates, it is necessary to evaluate its role in the specific scientific doctrines of the latter. Diderot's system of nature

141 "A quoy l'exemple de plusieurs corps, composez par l'artifice des hommes, m'a beaucoup servy: car je ne reconnois aucune difference entre les *machines que font les artisans* & les divers corps *que la nature seule compose.*" *Principes de la philosophie*, IV, 203; Descartes, IX, 321.

142 *ibid.*, XI, 37.

143 Diderot, II, 56. D'Holbach (*Système de la nature*, I, 11) conceived of nature in the same general sense: "La nature, dans sa signification la plus étendue, est le grand tout qui résulte de l'assemblage des différentes matieres, de leurs différentes combinaisons, & des différents mouvemens que nous voyons dans l'univers."

was in reality part of a much broader trend in natural science during the Enlightenment. Its immediate antecedents are to be found in the cosmogonical, geological, and biological speculations contained in two important works: Benoît de Maillet's *Telliamed,* and the preliminary portion of Buffon's *Histoire naturelle générale et particulière* published in 1749.[144]

The *Telliamed* set itself the task of outlining the natural process by means of which the earth in its present state, together with the various species of organic beings covering its surface, had been generated.[145] Maillet's treatment of the subject, it is true, made abundant use of fantasy; as a strictly "scientific" document its value is highly questionable. Yet, the effect that *Telliamed* had on the crystallization of naturalistic science among the philosophes could hardly be overlooked, or even exaggerated. First published posthumously in 1748, the treatise had been composed a decade earlier. Like the *Dissertation sur la formation du monde* in manuscript from 1738, it may be regarded as still another instance of that materialist current, with its interest in speculating on the world's origins from physical causes, that Pluche had in 1739 deplored so earnestly as being the outcome of Descartes's *Monde.*

The main importance of *Telliamed* lay, perhaps, in the fact that it drew forth and coordinated, by its insistence on the developmental viewpoint, certain scientific and philosophical tendencies. These latter, despite their presence in the general intellectual atmosphere, had remained, until Maillet's work appeared, belatedly inarticulate. The thesis of *Telliamed* began with the typically Cartesian postulate

144 Cf., on this point, my "From Deist to Atheist," *Diderot Studies,* pp. 58ff. Actually, the cosmogonical hypothesis of Buffon's *Théorie de la terre* was composed as early as 1744.

145 *Telliamed, ou Entretiens d'un Philosophe Indien avec un missionaire françois sur la Diminution de la Mer, la Formation de la Terre, l'Origine de l'Homme,* Amsterdam, 1748. The popularity of this book was such that it went through several editions in the next few years.

that "there is a perpetual motion in this universe, with regard to its very substance, and that a continual change is going on in all the globes of which it is composed."[146] In contrast with the ready-made world of Newtonian and Christian apologists, Maillet assumed that the earth—minor changes aside, which everyone was willing to admit—had not always been what it now seems to be. He attributed the formation of broad geological patterns to the gradual receding of the waters believed once to have covered the globe's entire surface. Consistently with this, the earliest version of a transformistic hypothesis was framed regarding the origins of animal life, which was supposed to have evolved, uniformly with the earth's development, from primitively aquatic forms.[147] The scientific background out of which *Telliamed* had grown is relatively easy to discern. Its frequent references to the *Pluralité des mondes* suggest its debt to Fontenelle's popular exposition of Cartesian physics. The book's being half-humorously dedicated to Cyrano de Bergerac recalls the naturalistic use already made of *tourbillon* physics in that writer's *Etats et Empires du Soleil.* But Maillet was still more definite. The general argument of *Telliamed* not only took for granted Descartes's vortex-theory, but made of it the all-inclusive physical basis for the vicissitudes assumed to occur constantly in the realm of nature.[148] The *Monde* underlay the conception central to *Telliamed* of a cosmos whose particular worlds were in process of successive destruction and reconstruction. Along with all this, Maillet was perfectly aware of the profound differences between his cosmo-biological speculations and the viewpoint of *Genesis.* After the usually ingenious but unconvincing attempt to affirm a harmony between the two accounts, *Telliamed* was obliged to declare, in agreement with Cartesian precept, that the spheres of natural science and theology

146 *ibid.*, I, XXX.
147 *ibid.*, II, 128ff.
148 *ibid.*, 68ff., 96-98, and *passim.*

were mutually independent and deserved to be treated as such.[149]

Not only were the ideas in *Telliamed* to comprise the core of Diderot's transformistic philosophy, but they were also promptly reflected, in much less fantastic form, in Buffon's *Théorie de la terre* of 1749, which played an indispensable part in promoting and consolidating the Enlightenment's materialist ideology. To explain the earth's geological formation, Buffon employed, like Maillet, the hypothesis of an original global sea. His theorizing suggested at the same time the rudiments of an evolutionary approach to organic phenomena, which Buffon first was to disavow under official pressure, then later reassert more firmly. Nonetheless, the *Théorie de la terre* offered to his early readers the composite picture of a world in process of development. The fuller context in which Buffon's geological and biological speculations occurred was, significantly, a cosmogonical system which, like Descartes's, sought to reconstruct the world with purely mechanical causes. In the "hypothetical section" of the *Théorie de la terre*, the origin of the planets was explained by the supposition that a comet, striking obliquely against the sun, had detached some pieces from its mass. For Buffon the earth was, therefore, at first a body of igneous, liquefied matter, which gradually cooled and grew opaque, and still retains at its center a vitreous substance. The earth's cooling caused the vapors originally surrounding it, until then in a state of rarefaction, to condense, thereby forming by their different densities the atmosphere and the initial global sea. It must be admitted that the *tourbillons* found no place in Buffon's cosmogony, for the translator of the *Méthode des Fluxions* was too well acquainted with Newtonian thought to put much stock in them. But what one does find in the *Théorie de la terre* is the author's determination, shared in common with the *Monde* and *Prin-*

149 *ibid.*, I, cxvii-cxviii.

cipes, to expound the formation of things fundamentally from matter in motion. The proof of this lies in Buffon's primary use of the "choc d'un comète," which was, in effect, the most radical feature of his general theory, and the notion that elicited the greatest amount of hostile comment from contemporary critics. The hypothesis of the comet's impact was intended to explain, on an intransigeantly mechanical basis, a phenomenon that Newton had been content to ascribe finalistically to God's will. This was, specifically, the fact that the planets rotate in their orbits around the sun in the same direction as the sun's rotation on its axis, for which the law of gravitation had been unable to assign a physical cause.[150] The mechanical impact of a comet striking the sun and plowing out fragments from its surface solved the problem in apparently non-teleological terms, and in conformity with the Cartesian requisite of *choc.*

To know the impression made by the *Théorie de la terre* on the Enlightenment, and particularly its relation to the materialist phase of Cartesian influence, the reactions of eighteenth-century readers ought to be consulted. Among these was Gautier d'Agoty, since forgotten, but at the time a respected colleague of Buffon's in natural science. What Gautier d'Agoty most vigorously impugned was Buffon's pretension to deal with the formation of the solar system without reference to final causes or theological traditions—a criticism to be echoed on all sides: ". . . to explain physically the manner in which the planets and their satellites have been formed is always, however it be presented, ridiculous, when it is not conceded that that is the work of the Creator. . . . He who believes to have discovered, by motion and the arrangement of matter (of matter such as it falls under our observation), how the order of the universe has been established, is in error; for that is a mystery hidden from philoso-

150 A. Hoffmann, "Die Lehre von der Bildung des Universums bei Descartes in ihrer geschichtlichen Bedeutung," *Archiv für Geschichte der Philosophie,* Bd. 17 (1903), p. 395.

phy."[151] Buffon, like Descartes, agreed that matter had, in a sense external to the scope of physical inquiry, been set metaphysically in motion by God. Notwithstanding, he proposed, again like Descartes, to disclose the process of development in nature without the intervention of final or intelligent factors. Despite all this, Buffon's cosmogony had no specific or necessary ties, any more than Descartes's, with a naturalistic metaphysics of the type that predicated the world's eternity, such as Hobbes, for example, had expressed. In the light of these considerations and Gautier d'Agoty's analysis, it is possible to understand the precise motivation and meaning of Buffon's naturalism, and its continuity with the background of Cartesian physics. The following passage, which brings out the contradiction posed by the "choc d'un comète" theory between naturalistic science and metaphysical providentialism, illustrates how the *Théorie de la terre* both drew upon and enriched the materialist current stemming from Cartesianism (Gautier d'Agoty's italics are quotations from Buffon):

"The Comet . . . is simply a hypothesis by means of which M. de Buffon has built his system about the formation of the earth and planets. It seems therefore, according to that author, that the earth is not eternal and that it has begun to exist. But what is the point of saying thereupon that *the force of impulsion has certainly been communicated to the stars generally by the hand of God, when it set the universe in motion; but* (there you have a 'but' that spoils everything) *since one must, as much as it is possible in Physics, abstain from having recourse to causes that are outside of*

[151] Gautier d'Agoty, *Observations sur l'histoire naturelle, sur la physique et sur la peinture,* Xe partie, Paris, 1754, p. 56. Against the Buffonian project, Gautier d'Agoty revoiced the usual *deus ex machina* approach to the formation of the planets: "Dieu se réserve cette connoissance à lui-même. Contentons-nous de connoître son existence dans les merveilles de la Nature, & n'allons pas substituer des Forces prétendues formatrices qui ne peuvent pas seulement être expliquées comme conservatrices des mouvemens." (p. 58)

Nature, it seems to me that in the solar system the force of impulsion can be accounted for in a sufficiently acceptable fashion, and that a cause for it can be found whose effects will be in accord with the laws of mechanics."[152]

The cosmogonical and geological speculations of *Telliamed* and the *Théorie de la terre* had had, of course, their immediate supports in a definite store of empirical data concerning, principally, the phenomena of rock-stratification and the fossilized remains of sea-life in land areas. One need, however, only consider how such facts were utilized by Buffon's chief predecessors to appreciate the originality of the *Théorie de la terre.* If Thomas Burnet's *Sacra Telluris Theoria* had as early as 1681 attempted a cosmogonical hypothesis, the emphasis had fallen heavily on the "sacra." Its preface declared the book to be "writ with a sincere intention to justifie the Doctrine of the *Universal Deluge*, and of *Paradise,* and to protect them from the Cavils of those that are no well-wishers to Sacred History." Similar in spirit and aims, except for replacing in the earth's origination Burnet's global deluge by the "atmosphere of a comet," William Whiston's *A New Theory of the Earth* (1696) had offered as scaffolding a lengthy exegesis of "the Mosaick Creation" before furnishing the specifically physical details overlooked by Scripture. And although John Woodward's *An Essay towards a Natural History of the Earth* (1695) had less to do with *Genesis*, because it investigated geological questions independently of any cosmogony, its author's desire was nonetheless "to assert the superintendence and Agency of Providence in the natural World: as also to evince the Fidelity and Exactness of the Mosaic Narrative of the Creation, and of the Deluge."[153]

152 *ibid.*, 59.

153 John Woodward, *An Essay towards a Natural History of the Earth, and Terrestrial bodyes, especially Minerals,* 3rd ed., London, 1723, Preface.

Inasmuch as the concern of the present study is, not with the history of cosmogony or geology as such, but with the manner in which those sciences contributed to the rise of naturalism, it becomes evident that Maillet's and Buffon's handling of facts and theories related to the pioneering efforts of Burnet, Whiston and Woodward actually invested those materials with a new significance. The Englishmen had purchased dearly the privilege of describing development in the natural order. Proceeding from a literal acceptance of a given interpretation of *Genesis*, their physical speculations, circumscribed and molded by theological tradition, were meant to give flesh to the Bible's skeletal account and to demonstrate the workings of Providence in the vicissitude of things. Ideologically considered, such preoccupations formed the background of the "physique de Moïse" with which the philosophes' critics hoped to frustrate the objectives of scientific naturalism. Buffon, fully cognizant of basic differences with the past, prefaced his cosmogony with circumstanced censure of the works of Burnet, Whiston, and Woodward, stressing their failure or inability to distinguish mechanical causes from the miraculous, physical evolution from Divine Creation, natural facts from Scriptural authorities—objecting, in short, to their serving up a *potpourri* of science and revealed religion. The same criticism was reproduced at length by Diderot in his *Encyclopédie* article on "Philosophie mosaïque et chrétienne," which saw in the efforts of pious cosmogonists a "monstrous mixture of theology with scientific systems, that has succeeded in degrading both religion and philosophy."[154] Making the division between

[154] Diderot, xvi, 122. Of Burnet's *Sacra Telluris Theoria*, "ouvrage où il propose de concilier Moïse avec les phénomènes," the encyclopedist said: "Jamais tant de recherches, tant d'érudition, tant de connaissances, d'esprit et de talent ne furent plus mal employés." *ibid.*, 127. Diderot cites, among many others, the amusing instance of one master in edification who offered to give "une grammaire, une rhétorique, une logique, une arithmétique, une géométrie, une optique et une musique chrétiennes. Voilà les extravagances où l'on est conduit par un zèle aveugle de tout christianiser." *ibid.*, 125.

questions of faith and of natural science absolutely irrevocable, Diderot's purpose was to affirm that "la nature est le seul livre du philosophe." In contrast with what Buffon and Diderot reproved, the *Théorie de la terre*, like the *Telliamed*, was resolved to deal with cosmic origins wholly apart from any theological presupposition or exigency, and solely with the known or knowable mechanical properties of matter.

Towards the mid-century in France, the stream of cosmo-geological theories flowing from various sources merged with the peculiar current of naturalism ensuing from Descartes's proposal to construct a world with matter in motion. Nothing, moreover, could have been more normal and logically appropriate than this merger. For not only was Cartesian-inspired materialism capable of providing, during the Enlightenment, an ideal *rationale* for the program of exploration in geology and biology, but the acquisitions of cosmo-geological inquiry itself, historically considered, were closely affiliated with the broad influence of Cartesianism. Long before the appearance of the *Théorie de la terre*, Fontenelle had commented advisedly on just that point: "Descartes was first, for it often happens that the history of some particular research or of some discovery begins with him, who had the idea of explaining mechanically the earth's formation. Thereafter, Steno, Burnet, Woodward, Scheuchzer have taken over, extended, or rectified his ideas, and have combined them variously with one another."[155] The separate trends of cosmo-geological science on the one hand, and of philosophical naturalism on the other, having issued from a common source in Cartesian thought and followed, as it were, the paths of two intersecting arcs, could be expected,

155 *Histoire de l'Académie royale des Sciences*, 1708, p. 30. The occasion for this comment was the communication of a memoir by the Swiss naturalist, Jean Scheuchzer, in which he discoursed on the earth's formation in the vein of Woodward and the "diluvian school," hopelessly embroiling physics and theology.

sooner or later, to converge and unite again. Their convergence around 1750, representing a reintegration of the Cartesian heritage of physical doctrine, was one of the principal immediate causes behind the emergence of scientific naturalism among the philosophes. The cosmogonies of Maillet and Buffon were to furnish, moreover, the hypothetical setting for certain speculations in biology based on a transformistic view of organism; but this subject belongs to another chapter.

Meanwhile, the roots of Diderot's natural philosophy, as of the materialism it occasioned, are revealed. In the *Lettre sur les aveugles*, published a year after *Telliamed* and in the same year as the *Théorie de la terre*, Diderot gave a brief but crucial *ébauche* of his future system of nature. He reiterated Maillet's evolutionist conception of the origins of life, and contemplated physical phenomena uniformly from the standpoint of developmental process. The naturalism of Saunderson's deathbed discourse, expressed as a retort to the attitude of a disciple of Newtonian theology, was intimately related to the transformist interpretations or innuendos present in *Telliamed* and the *Théorie de la terre.* The spokesman for Diderot told his protagonist that Newton's chief error lay in the gratuitous assumption that things had always been what they now are. Notwithstanding the present order in the cosmos, there was for Saunderson nothing either final or finalistic about the patterns of nature:

"I submit to you concerning the actual state of the universe, in order to obtain from you in return the liberty of thinking what I wish about its first and primitive state. . . . Imagine therefore, if you like, that the order which strikes you has always existed; but allow me to believe that such is not the case at all; and that if we were to go back to the birth of things and of time, and if we had experience of matter moving itself and chaos arranging itself into order, we would meet with a multitude of formless beings as against a few well-organized ones. . . ."[156]

"What is this world . . . ? A composite subject to revolutions, all of which indicate a continual tendency towards destruction; a rapid succession of beings that follow one another, push each against the other and disappear; an ephemeral symmetry, a momentaneous order."[157]

Enlarging upon these thoughts, and acquiring a close knowledge of Buffon's *Histoire naturelle* which, while imprisoned for the *Lettre sur les aveugles*, Diderot had found leisure to study and comment,[158] the author of the *Interprétation de la nature* of 1754 elaborated further the biological consequences of Saunderson's views, in particular as they bore upon a vitalist, transformist theory of organic origins. That Diderot's natural philosophy was in the main stream of Descartes's initial attempt to construct a world from matter in motion was noted, among others, by Gautier d'Agoty, who saw it as a kind of logical continuation of the *Théorie de la terre*: "I place behind the Comet of M. de Buffon, the Hypothesis of M. Diderot, because I find much correspondence between these two systems."[159] The theories of Buffon and Diderot complemented one another, according to the critic of scientific naturalism, by their common aim to found a science dealing with development that would be at once mechanistic and non-finalistic: "We have sufficiently criticized that imaginary force of the comet which, without

[156] Diderot, *Lettre sur les aveugles*, pp. 41-42.

[157] *ibid.*, 44. D'Holbach too dwelt upon the vicissitudes of a cosmos which, like that of Descartes, was subject to successive generations and destructions: "Tout est en mouvement dans l'univers. . . . Tout ce qui nous semble en repos ne reste pourtant pas un instant au même état: tous les êtres ne font continuellement que naître, s'accroître, décroître & se dissiper avec plus ou moins de lenteur ou de rapidité." *Système de la nature*, I, 18-19; see also, I, 32-33.

[158] Diderot, XIX, 422-423.

[159] Gautier d'Agoty, *Observations sur l'histoire naturelle*, Xe partie, p. 58n.

reference to *causes that are outside Nature*, has moved of itself and has by its mechanical impact brought about the separation of solar matter, from which the earth and planets have produced themselves. Now a friend of that philosopher's, M. Diderot, has added to this that there is in the world a *living matter* and an *inert matter* from which all beings have formed themselves."[160] Gautier d'Agoty defined, as well as anyone could at the time, the purpose and meaning of scientific naturalism: "Who would not believe, on reading about the marvelous effects of the comet and the beginning of M. Diderot's book [the *Interprétation de la nature*], that this author and M. de Buffon wish concertedly to establish a science both natural and purely mechanical?"

Diderot's system of nature grew, then, out of the materialist trend for which the abbé Pluche had held Descartes's *Monde* responsible. In a diatribe against Diderot by the abbé Sennemaud, the systems of "our modern Democrituses" were said to have resulted from a type of *incrédulité* which was, seemingly, another word for what Pluche had described: "Physics owes to it those daring and sublime hypotheses that have taught us five or six different manners of creating the world, and of setting in operation the numberless creatures that make it up."[161] On the basis of the variety of evidence thus far adduced, the parallelism between the Cartesian *Monde* and Diderot's natural philosophy may be stated succinctly in its essential character. The *Monde* had had its inception in a primordial Chaos, "the most confused and embroiled that poets can describe"; but given the laws of moving matter, they had sufficed to cause "the particles to disentangle from one another by themselves, and to arrange themselves in such an order as to give rise to a very perfect

160 *ibid.*, 59.

161 Abbé Pierre Sennemaud, *Pensées philosophiques d'un citoyen de Montmartre*, La Haye, 1756, p. 63. Although directed principally against the *Pensées philosophiques* of 1746, Sennemaud's polemical satire alluded also to ideas in Diderot's subsequent writings.

world."[162] Likewise, in the *Lettre sur les aveugles*, Diderot placed himself fundamentally at the Cartesian point of departure, where "matter in fermentation caused the world to take shape."[163] His general orientation in natural science was fixed, like Descartes's, by the hypothetical vision of "matter moving itself and chaos arranging itself into order." In such a conception, Chaos did not mean the absolute disorder whose transfigurations both Christians and Epicureans attributed to an inexplicable cause. For Descartes and Diderot, Chaos was the entirely indeterminate state of matter whose innumerable potential forms, realizable successively in time, were determined by a mechanical causation inherent to it. Transcribed to the logical level, it became, in the total posing of the problem of cosmic origins, the theoretical starting-point requisite for a purely intelligible solution.

Diderot generalized and applied to the cosmos as a whole the transformist principles assumed, more specifically, to explain the production of organic beings. Such a step was in keeping with the wellworn cosmogonical theme that owed its career historically to the *tourbillon* astronomy: "Why should I not extend to the universe what I believe to be true for animal life? How many imperfect mangled worlds have already perished, or are taking shape again and disappearing perhaps at every moment somewhere in distant space . . . where motion continues and will continue to combine masses of matter until they attain an arrangement in which they will persist?"[164] Diderot's *Principes philosophiques sur la matière et le mouvement*, by rejecting the definition of matter as a homogeneous substance indifferent to motion or repose, actually represented the triumph of Descartes's physical over his metaphysical principles. In other words, Diderot elaborated the implications, particularly for biology, of the Cartesian reduction of all natural patterns to matter in

[162] Descartes, XI, 34-35.
[163] Diderot, *Lettre sur les aveugles*, p. 43. [164] *ibid.*

motion exclusively. But if the transformistic vitalism of the *Rêve de D'Alembert* supposed that matter was necessarily endowed, not only with energy, but with *sensibilité*, it will appear later that this quality, however un-Cartesian at first glance, proved analytically undistinguishable, as the *Entretien entre D'Alembert et Diderot* understood it, from mechanical cause.[165]

Whatever the specific differences were between Diderot's panvitalist system and Descartes's *Monde*, the problem here is to ascertain how Cartesian physics fostered that special ideology in the context of which the investigations of Diderot's group occurred and flourished. That Diderot recognized such an influence by the *Monde*, and agreed with Descartes's basic method and aims in that work, may be surmised from a story about the philosophe told by a near associate. In 1765, on the occasion of the French Academy's offer of a prize for the best eulogy on Descartes, Thomas had sought Diderot's opinion—perhaps for good reasons—concerning Cartesian physics. The philosophe's reply, summarizing the *Monde* and *Principes*, shows that, while he had use neither for vortices nor for the particular laws of motion to which they conformed, his interest centered sympathetically in Descartes's ambition to deduce the actual state of the universe solely from matter in motion:

". . . one day Descartes said to the Eternal Being: Give me matter and motion, and I too will create a world. And the Eternal One gave him matter and motion and said: Let us see how the atoms will behave in order to give rise to a world. And Descartes ordered matter to move with a circular motion, and decreed that its component parts should submit themselves to the laws of bodies in circular movement. The Eternal One said with astonishment: That is as I have done. And smiling, He applauded the philosopher. But when He

[165] Diderot, II, 106.

saw him giving in to his imagination, substituting chimerical notions for the properties of bodies and their eternal laws, and going astray amid his vortices, the Eternal Being turned away and reposed."[166]

Diderot's *Encyclopédie* article on "Chaos" introduces us directly into the ideological issues and historical background of its author's naturalism. Defending, with tongue in cheek, the orthodox position in cosmogony against impious incursions, Diderot comes forthwith to the heart of the question. To imagine, he says, "à l'exemple de quelques systématiques," and contrary to the Mosaic Creation, that God produced in the beginning merely an undefined, undetermined material substance, which by means of motion and various mechanical processes gradually became a world; or to embrace a less ambitious theory, such as Whiston's, concerning the planets' origins, is to accredit improbable opinions in place of the eternal truths that God has revealed by the mouth of Moses. If made by a theologian or even a deist at the time, the criticism would sound genuine and solemn enough. Coming from Diderot, its irony is thinly veiled and becomes more apparent with insinuation:

"This physics of Moses, which describes the Eternal Wisdom regulating nature and the functions of each thing by so many express decisions and commands; a physics that makes reference to general, constant, and uniform laws only to maintain the world in its one and only state, and not to form it, is certainly worth the imaginings of systematists, be they the classical materialists, or the modern physicists

[166] Grimm, Diderot, Raynal, Meister, *et al., Correspondance littéraire, philosophique et critique*, VI, 354. This anecdote is in all likelihood a faithful paraphrase of what Diderot must actually have said, for its author adds quite seriously that Thomas should have used the complete account, not merely part of it, in his prize-winning *Eloge*. There is no reason to assume that the *Correspondance littéraire* would deliberately have attributed spurious or improbable opinions to its own editor. Incidentally, Diderot thought very highly of Thomas' oration. XIX, 154.

who derive all beings from a homogeneous matter moved in all directions."[167]

The concluding phrase would seem to allude to Descartes's doctrine, as distinct, for example, from the theological adaptations of the "static" Newtonian physics. Diderot further remarks that such a science of nature would tend towards Spinozism, i.e. atheistic materialism, in eighteenth-century parlance. But if any doubt remains about the identity of the "modern physicists" whose teaching, by infusing new life into Epicureanism, has brought about the ideology that Diderot himself best represented, this is dispelled without the shadow of equivocation:

"One cannot help but remark here how unsure philosophy is in its principles and unpredictable in its behavior. It was claimed once that movement and matter were the only necessary things. If, since then, philosophy persisted in maintaining that matter is uncreated, at least it was subjected to an intelligent Being in order to make it assume a thousand different forms, and to dispose its particles in that appropriate arrangement from which the world springs. Today it is admitted that matter is created, and that God communicates motion to it; but it is asserted that this motion, emanating from the hand of God, is able, when left to itself, to account for all the phenomena of the visible world. A philosopher who dares undertake to explain, by the laws of motion alone, the mechanism and even the original formation of things, and who says: *give me matter and motion, and I will make a world*, ought first to prove (which is quite easy) that existence and movement are not at all essential to matter. Otherwise, that philosopher, believing erroneously that there is nothing in the marvelous spectacle of this universe which motion alone could not have produced, is in danger of falling into atheism."[168]

[167] Diderot, xiv, 90-91. [168] *ibid.*, 91.

Diderot, then, had perfect knowledge of the radical possibilities of Descartes's physics taken separately from the metaphysics. His words bear out, moreover, the thesis that the *Monde* and *Principes*, rather than the contributions either of Gassendi or Hobbes, were the decisive source of the Enlightenment's scientific naturalism. A typical manifestation of this general trend may be seen, also, in the publication in 1751 of a work such as Mirabaud's *Le Monde, son origine et son antiquité*. Written much earlier and restricting itself merely to a historical exposition of Graeco-Roman cosmogonies, the timeliness of the book's appearance in print did not escape the notice of contemporaries.[169] The main interest of Mirabaud's somewhat recondite discussions lay, as everyone realized, in the support it lent indirectly to the theories already voiced by Maillet, Buffon, and Diderot.

Diderot himself, it would seem, discreetly ascribed to Epicurus, on occasion, what he had received in all likelihood from the materialist phase of Cartesianism. In his *Encyclopédie* article "Epicuréisme," for instance, the following lines might well be taken to reflect, not so much what Gassendi had taught, as what Descartes's science had adopted and reinterpreted from the storehouse of Epicurean and Renaissance ideas: "The plurality of worlds is not an absurd notion. It is possible that there are other worlds like ours, as well as different ones. One must conceive of them as large *tourbillons* supported each against the other, and between which are compressed smaller *tourbillons*, that all together fill up the infinity of space. Amidst the general movement that produced our world, the mass of atoms which we call the *earth* occu-

169 The abbé Guidi's *Entretiens philosophiques sur la religion* (Paris, 1771) attacked at length Mirabaud's work for being representative of the materialism of the age. Gauchat's *Lettres critiques* (xv, 103-111) accurately pointed out that a long passage interpolated in the book, and dealing with the supposition of a primordial deluge, had been plagiarized *verbatim* from the *Telliamed.* Concerning Mirabaud and his writings, see Wade, *Clandestine Organization and Diffusion of Philosophic Ideas*, pp. 205-221.

pied the center; other masses went to form the heavens, and the stars that illuminate them."[170] Some years later, the catechism of the materialist school appeared, rather belatedly: D'Holbach's *Système de la nature.* D'Holbach linked the principles underlying his argumentation with what he understood to be those of Descartes's natural philosophy:

". . . in supposing . . . the existence of matter, one must suppose that it possesses certain qualities from which its motions or modes of behavior, determined by these same qualities, must necessarily follow. In order to form the universe, Descartes asked for only matter and movement. A diversified matter was sufficient for him; its different motions were the consequences of its existence, its essence and its properties; its various modes of action are the necessary results of its different modes of being. A material substance without properties is nothing but pure non-being. Thus, from the moment that matter exists, it must act. . . ."[171]

In conclusion, a sampling of comments from the contemporaries themselves of the philosophes' scientific naturalism will corroborate the viewpoint of the present chapter. Their observations do not, as a rule, refer directly to Diderot.

170 Diderot, xiv, 515.

171 D'Holbach, *Système de la nature*, i, 30. The postulate of "motion essential to matter" was of course found by the philosophes in John Toland. But when in 1768 D'Holbach published a translation of the *Letters to Serena* under the title of *Lettres philosophiques*, the influence of Descartes may be said, by a curious circuit, to have reached the French Enlightenment by way of England. Toland's idea had been a consequence of both the Cartesian definition of matter as extension and the related *plenum* theory; as such, it had expressed itself within the specific framework of the *tourbillon* system. Cf., *Letters to Serena*, London, 1704, pp. 186-192. With this, Toland's "motion essential to matter" was reasoned out in conjunction with the basic view of a vortex universe in perpetual flux and development: "All the Parts of the Universe are in this constant Motion of destroying and begetting, of begetting and destroying: and the greater Systems are acknowledg'd to have their ceaseless Movements as well as the smallest Particles, the very central Globes of the Vortexes turning about their own Axis; and every Particle in the Vortex gravitating towards the Center." *ibid.*, 188.

The finished form of Diderot's system of nature, as expressed in the *Rêve de D'Alembert*, was of course not published until 1830. What appeared in his lifetime, in the *Lettre sur les aveugles* and the *Interprétation de la nature*, did not, moreover, profoundly trouble the adversaries of materialism. The reason was that Diderot's earlier writings gave his natural philosophy in a somewhat fragmentary and disconnected form not readily intelligible to the reading public; also, complexity of meaning made it rather esoteric. As a result, the critics of naturalism understandably chose to train their weapons on such widely-read, influential writers as Benoît de Maillet, Buffon, and D'Holbach. But what was said about these may for the most part justly be applied, in the interest of historical reconstruction, to Diderot as well.

Among the philosophes themselves, Voltaire especially was aware of the relationship that Descartes's *Monde* bore to the materialist science towards which, as a resolute Newtonian deist, he remained implacably hostile. Alluding apparently to such figures as Diderot, Buffon, and D'Holbach, Voltaire remarked: "More than one philosopher . . . acting on the example of Descartes, has wished to put himself in the place of God and to create, like Him, a world by means of words; but soon all these follies of philosophy are reproved by the wise."[172] The *Dissertation sur les changements arrivés dans notre globe* acknowledged, moreover, the part played by the cosmogonies of *Telliamed* and the *Théorie de la terre* in bringing certain Cartesian ideas to maturity and eventual triumph. On the geological theory of a primitive global sea, Voltaire reflected ironically: "A longer time than the duration of the deluge would be required to read all the authors who have erected handsome systems by means of it. Each destroys and rebuilds the earth in his fashion, just as Descartes fabricated it; for the majority of philosophers have put themselves without ceremony in God's place, and think

[172] *Précis du siècle de Louis XV*, 1768; Voltaire, *Œuvres*, xv, 433-434.

of creating a universe by use of words."[173] In a letter to Buffon, Turgot, like Voltaire, made known his feeling about the basic affinity between Descartes's *Monde* and the project, in the *Théorie de la terre*, to elucidate the origins of the solar system: "I ask, first, why do you undertake to explain such phenomena? Do you wish to deprive Newtonian science of that simplicity and wise restraint which characterize it? By plunging us back into the obscurity of hypotheses, do you want to give justification to the Cartesians, with their three elements and their formation of the world?"[174]

The critics of Enlightenment went more deeply into the problem. Descartes's contribution to the naturalism of the age was plainly designated and analyzed as such in Denesle's *Examen du matérialisme* of 1754. Intended as an antidote for the spread of the ideology typical of Diderot's coterie, this book had a bone to pick with the supposed remark of the "Philosophe conséquent & géometrique" to the effect: "Give me matter and motion, and I'll make you a universe." Denesle argued that the dualism of Cartesian metaphysics had been undermined completely by Descartes's own principles in physics: "one can dispense with recognizing . . . two essentially distinct substances if matter, once set in motion, is able to suffice for everything."[175]

Glancing at a work by the abbé Dufour some years later, which likewise opposed the materialist heresy of the day, one may note the same historical filiations. In its pages, the philosophes' naturalism is traced unmistakably to its sources

[173] *ibid.*, xxiii, 226. In Maillet, Buffon, and Diderot, the theory of transformism applicable to the emergence of organic beings was asserted in conformity with the notion that the whole of the cosmos had itself evolved. Voltaire, by an inverse reasoning, sought to establish instead the static Newtonian world-order by making it the *a posteriori* result of the assumption, held in common by eighteenth-century apologists and compilers of natural theologies, that the various species of vegetative and animal life were themselves constant and invariable: "il serait bien étrange que la graine de millet conservât éternellement sa nature, et que le globe entier variât la sienne." *ibid.*, 228.

[174] Turgot, *Œuvres*, ii, 96.

[175] Denesle, *Examen du matérialisme*, i, 35.

in Cartesian natural science. Dufour observes astutely that it has made little difference whether matter was defined metaphysically as finite or infinite (a point that Descartes tried to settle by making it "indefinite"). The attitude crucial for materialism's progress has been the determination to have all things proceed from matter, which is in turn conceived to be "as subtle or as gross as might be necessary in order to explain everything."[176] Such a scheme of explanation, according to Dufour, went hand in glove with the vortex hypothesis and, in general, with its naturalistic tendencies. After summarizing the meaning of vortex physics, the critic is unable to accept the supposed passivity of matter which, in Descartes's metaphysics, required the agency of God to be set in motion. We are indeed not far from the thesis of Diderot's *Principes philosophiques sur la matière et le mouvement*, when Dufour concludes: "It would be ridiculous to imagine that the *tourbillons* which compose the universe, being in a state of continual movement, should ever remain in repose. Everything in nature moves; nothing is in repose . . . an extremely subtle matter penetrates all, moves all, animates the less subtle matter. . . ."[177] Thus understood, Cartesian physics was obviously related to that developmental view of nature which cleared the path for the philosophes' scientific naturalism: "Such is the effect of perpetual motion: it forms and destroys, only to re-form in another manner. Without altering the substance of nature, it continually changes its outward appearance."[178] The materialist school's definition of Nature, as analyzed by Dufour, corresponds fundamentally with the general notions of Descartes and Diderot already referred to in this chapter. Nature is "nothing but the order observed in its course of progress by the universal movement, and the expression of that order"; or more pre-

[176] Abbé Dufour, *l'Ame ou le sisteme des Matérialistes, soumis aux seules lumières de la raison*, Avignon, 1759, p. 8.
[177] *ibid.*, 11. [178] *ibid.*, 12.

cisely, "the action of the components of matter one upon the other, and the ordinary effect of movement in the formation, conservation, and destruction of all beings. The idea of nature coincides, in itself, with that of motion."[179]

The critical issue, therefore, was: did movement actually inhere in matter, or was it essentially apart from it? In the Enlightenment, the divergent orientations towards scientific naturalism or towards natural theology depended on how this question was resolved. A certain Boullier, examining the problem, favored the apologetic finalist solution, as opposed to the standpoint of materialists. He maintained that motion ought to be considered a mode positively imprinted on matter by the agent who moves it. Of special interest to us is the fact that Boullier was led, obviously in view of the evolution of Cartesian physics, to state that Descartes, in the *Principes*, had taken the opposite position and claimed that motion was a mode really belonging to bodies.[180] Boullier felt that Descartes could not have been serious. Be this as it may, the debt of eighteenth-century materialists to Cartesian precedent was quite serious.

The opinion of Gauchat's *Lettres critiques*, linking the naturalism of the Enlightenment with the influence of Descartes's science, has been mentioned at the start of this chapter. Particularly significant is the circumstance that, although the abbé wrote after the culmination-period of the materialist current in the 1750's, his estimate of Descartes's heavy responsibility did not differ substantially from what Pluche had said, a score of years earlier, in anticipation of what was to happen. In 1761, Gauchat took the collaboration between Cartesian thought and the philosophes' naturalism to be more or less a *fait accompli*. His criticism of the

179 *ibid.*, 25-26.

180 David-Renaud Boullier, *Discours philosophiques: sur les causes finales; sur l'inertie de la matiere; sur la liberté des actions humaines*, Amsterdam, 1759, p. 92. The reference to Descartes is in the *Principes*, ii, 29-30.

latter was limited to an examination of *Telliamed* and Buffon's cosmogony. In Gauchat's evaluation, Descartes's unfinalistic manufacture of a world had enclosed two pernicious possibilities, of use to materialism. The first was that it attributed to matter a force which it does not possess, and "even a sort of wisdom."[181] To claim that matter once set in motion could result in the actual arrangement of nature without aid of a divine blueprint, was tantamount to "bestowing upon it essential properties of force and order; it means that all matter in action produces regular, organized, and sensitive bodies."[182] The logical outcome of this was plainly the geological and biological speculations of *Telliamed* and the *Théorie de la terre.* Gauchat himself believed, however, that Descartes's corpuscular matter and specific laws of motion could no more have engendered an ordered cosmos "than the movement in all directions, and to infinity, of Leucippus' atoms." The fact that Descartes's physical principles differed from those of Leucippus did not, for Gauchat, greatly affect the underlying agreement between the two philosophers with respect to materialism; since, once their differing types of motion were granted, God added no more to Cartesian physics than to the Leucippean.[183]

181 Gauchat, *Lettres critiques,* xv, 266. All but forgotten today, the abbé Gauchat composed between 1755 and 1763 a nineteen-volume refutation of the multiple aspects of Enlightenment, and was eminently qualified to speak on the sources and meaning of the period's materialism.

182 *ibid.*, 268.

183 *ibid.* As regarded the sciences of development, Gauchat wished to halt all inquiry at the stage of *Genesis.* He held that cosmogonical and geological speculations explaining the age of the globe and its formation in contradiction to Christian chronology and the Mosaic "physics" were, in effect, attacking religion at its roots (pp. 66-67). In the same critical spirit, the abbé Nonnotte gave a revealing classification of names: "As for the romances of Maillet, Buffon, Descartes, and Epicurus, which see nothing but motion in the forming of the world . . . how pitiful these tales are compared to what the Doctor of the Hebrews offers us!" *Dictionnaire philosophique de la religion,* II, 41. Opposed to such narrow views of the problem of natural science, D'Alembert's *De l'Abus de la critique en matière de religion,* epitomizing Descartes's physics by the current dictum: "Give me matter and motion, and I'll make a world,"

Descartes's second error, in Gauchat's eyes, was that he had ascribed to God a strictly metaphysical, but not a physical, relationship to nature. True, God had set matter in motion and constantly preserved its essence; but the causation of particular effects in nature was referred, not to God, but to the laws of matter in motion. Hence, it ensued from Cartesian doctrine that in the domain proper to physics theology could hope to find neither support nor application of any kind. As regarded God's preservation of the essence of material substance, Gauchat pointed out that this, too, had abetted rather than hindered the materialist career of Descartes's science: "The decisive point, the source of error (it cannot be too often repeated) is that two fundamentally different things have been confused: the creation of the world, and its conservation."[184] Descartes, it would seem, had inverted the correct relationship of these to God, making God responsible for the preservation of the universe *quâ* matter, but matter in motion the sole cause of its creation. Gauchat observed that Descartes ought to have attributed, inversely, the creation to God, and the regularity and constancy of cosmic functions, as Newtonians did, to the universal laws of mechanics. From the dangerous precept of Cartesianism, according to the *Lettres critiques*, there had arisen such scientific endeavors as *Telliamed* and the *Théorie de la terre* which, by their determination to expound the origin of things by matter in motion only, laid the foundations of materialism.

In his critique of Buffon's cosmogony, Gautier d'Agoty had already remarked about Descartes's version of the Prime Mover argument, that it was a mistake to have God

justified it against the bigotry of those "enemies of reason" who saw in it the materialism or atheism of an "arrogant fabricator of systems." D'Alembert agreed with Diderot in vindicating the scientist's right to give a physical account of the origin of things without interference from theology. *Œuvres*, I, 551-552, 558-560.

[184] Gauchat, *op. cit.*, XV, 269-270, and ff.

grant motion to matter and thereupon to abandon it to itself; for an impious mind could suspect that movement had always existed, if it was able to perpetuate itself without the active help of the Deity. Gautier d'Agoty added: "And it would suffice to prove that it had existed in this manner for a single moment in order to prove that it had existed so from all eternity: there you have the mask with which the materialists who live among us have covered themselves."[185] In 1738, Voltaire had described with unusual perspicacity the transition from natural theology to naturalism implied by Cartesian physics: "As soon as one is persuaded, with Descartes, that the world cannot be finite, and that motion exists always in the same quantity; when one dares to say, Give me movement and matter and I will make a world; then it must be admitted that these notions exclude, by consequences that are quite justified, the idea of a single infinite Being who is the sole author of motion, and the exclusive cause for the organization of substances."[186] The career and content of eighteenth-century materialism, and Diderot's leading role in it, bore out the accuracy of Voltaire's judgment.

[185] Gautier d'Agoty, *Observations sur l'histoire naturelle*, XIe partie, 1754, p. 17.
[186] Voltaire, *Œuvres*, XXII, 404.

CHAPTER III

SCIENTIFIC METHOD FROM DESCARTES TO THE PHILOSOPHES

Quand la curiosité se trouve jointe à un peu d'imagination, on veut aussitôt porter la vue au loin, on veut tout embrasser, tout connoître. Dans ce dessein, on néglige les détails, les choses à notre portée; on vole dans des pays inconnus, et on bâtit des systêmes.

—Condillac, *Traité des Systèmes*

CHAPTER III

SCIENTIFIC METHOD FROM DESCARTES TO THE PHILOSOPHES

THE scientific naturalism of the philosophes has been traced to their endeavor, under the aegis of Descartes's physics, to account for the original organization of the natural order in terms strictly of matter in motion. The execution of such a plan could hardly have been divorced, however, from the methodology in science that the Enlightenment had also inherited from the same general source. In reality, Cartesian tradition was to act from two distinct but converging directions on the growth of the ideology represented by Diderot and his colleagues. The first of these was the impact, both cumulative and progressive, of the diversified materialist-pantheist current which, issuing from the *Monde* and *Principes* and fed by a number of tributaries, had gained momentum up to the 1740's. It finally culminated, logically and chronologically, in the cosmogonical, geological, and biological speculations of Buffon, Diderot, and their followers. The second influence, that of Descartes's methodology, came to the philosophes in the less tangible form of an all-pervasive trait of the eighteenth-century French mind. But the continuity, such as it was, from Descartes to the materialist school both in scientific method and in physical principles was owing, ultimately, to the common goal of deducing the formation of the world's manifold patterns from a definition of matter and its mechanical laws.

The linking of Descartes and Diderot on methodological grounds involves an apparent difficulty. It is recalled from the start that the philosophes, adopting the experimentalism of Bacon and Newton, were monotonous in their condemnations of Descartes's ideal of *a priori*, rationalist science, as of his "esprit de système." Condillac's *Traité des systèmes* of 1749 became the well-thumbed manual of the Enlightenment's anti-systematic and pro-experimentalist phase, at the same time that vortex physics, under attack by the philosophes, was in the throes of a slow but well-merited death. Conversely with the decline of Cartesian science, Newton's physics and methodology triumphed in France and became, through the concerted efforts of Voltaire, Maupertuis, D'Alembert, and others, in some manner wedded to the cause of Enlightenment and progress itself. In this broad trend, Diderot himself was reputed the most advanced exponent of Baconian experimentalism.

Such is, in brief, the state of affairs that scholarly opinion, reaching back more than a century and typified by a reluctance to probe beneath the surface, has pictured. If not exactly untrue, the picture is at least misleadingly incomplete. Beyond doubt, the introduction of English experimental procedure into France was one of the philosophes' major accomplishments, and its effect on their thinking, when and where it found application, was both profound and lasting. But that is not the question here. The methodological tenets derived from Bacon and Newton, it will appear, were fundamentally irrelevant, and often even antagonistic, to the general ends proposed specifically by scientific naturalism. In this vast area of eighteenth-century thought, the influence of Cartesian methodology, actively persisting side by side with the Newtonian, proved to be the dominantly operative factor.

On this point Diderot himself, the foremost *méthodologue* of the age, is the best authority to consult. His own philosophy of nature, the most penetrating and suggestive during the Enlightenment, was patently not an instance of the experimentalist credo set forth in the greater part of the *Pensées sur l'interprétation de la nature.* Those scholars who have exaggerated Diderot's loyalty to Baconian teaching quite out of proportion to its place in his thought as a whole,[1] have been confronted with a serious inconsistency. In the actual elaboration of his scientific doctrine, the encyclopedist was only too often given to bold speculation, sweeping hypotheses, and to a deductive reasoning related rather to the aprioristic penchant of his imagination than to any exigencies of observed fact. Critical appraisals of this seeming contradiction, ordinarily stressing Diderot's love of paradox and his tendency to theorize with his passions or temperamental enthusiasm, not only have been unjust to the philosophe as thinker, but have neglected certain Cartesian elements in the conception of science prevalent during the Enlightenment. Consequently, the present chapter will seek, first, to define the general sense in which Descartes's scientific method received concrete expression in eighteenth-century France. Secondly, the extent of its more special contribution to Diderot's methodological theory and practice alike will be examined. This will in turn permit, finally, a demonstration of its intimate ties with the speculations in natural science that formed the basis of the period's materialism.

Throughout the Enlightenment in France, two distinct orientations pertaining to scientific method were in force, even intermingling frequently in varying combinations. One was the experimental, inductive procedure as it had developed through Galileo, Bacon, Pascal, Boyle, Newton; the other was the *a priori*, deductive attitude that had stemmed from Cartesian philosophy.[2] The history of scientific method among the philosophes could well be written in terms of the interactions between these two divergent and controlling views, which almost everyone at the time habitually desig-

1 A few of the numerous instances are: R. L. Cru, *Diderot and English Thought*, New York, 1913; F. Venturi, *Jeunesse de Diderot*, Paris, 1939, esp. pp. 286-293, 296, 313-314; J. Luc, *Diderot*, Paris, 1938, p. 107; D. Mornet, *Diderot, l'homme et l'œuvre*, Paris, 1941, pp. 41ff.

2 Cf., on this question, Charles Frankel, *The Faith of Reason*, New York, 1948, esp. pp. 3-5, 13-38.

nated as the "Newtonian" and the "Cartesian." The Newtonian-Cartesian conflict, in physics as in methodology, was thus a polarity historically peculiar to the philosophes' *milieu.* Apart from it, Diderot's reflections on scientific method, typically oscillating between two extremes, could not be correctly understood. In the *Pensées sur l'interprétation de la nature,* for example, the dichotomy between "philosophie rationelle" and "philosophie expérimentale," announced from the start and running through the work, would be, without regard to this polar situation, a remote and puzzling generality rather than, as it was, a conscious preoccupation with the contemporary problem of a science of nature. As his ultimate objective in the *Interprétation de la nature,* Diderot sought to harmonize the two opposites. What that synthesis was, and how it complemented the formulation of his materialism—which also figured in the pages of the same book—can best be stated only after the separate roles of experimentalism and of rationalism in Diderot's natural philosophy have been assigned each its proper value.

Writing to Maupertuis in 1738, Voltaire gave his private estimate of the stage that the "Descartes *versus* Newton" polemic had then reached: "I see that thinking in France is at a point of great ferment, and that the names of Descartes and Newton seem to be the rallying-calls of two parties."[3] The controversy to which Voltaire referred had, however, gradually taken shape in the course of several decades. The progressive infiltration of Newtonian physics into France up to 1738 had, from the first, evoked a tenacious resistance from the majority of the members of the *Académie des sciences.* They had remained in an official manner, even if yielding on many particular issues, attached to Cartesianism.

[3] Voltaire, *Œuvres,* xxxv, 2. A certain abbé Dambésieux, siding with Descartes, was still able to speak a score of years later of "deux systêmes qui partagent aujourd'hui le monde pensant." *Reflexions sur la physique moderne; ou, la philosophie newtonienne comparée avec celle de Descartes,* Paris, 1757, Avant-propos, xiii.

The continuous discussions that took place during that long period in favor either of Newtonian or of Cartesian ideas, and which resulted in the eventual abandonment of the *tourbillons* and related notions, constituted the historical background of the special polarity that the philosophes inherited.[4]

Although Descartes's physics itself began to be discarded, as D'Alembert remarked,[5] some thirty years before the first volume of the *Encyclopédie* appeared in 1751, the scientific methodology that had inspired it was to fare more fortunately. Even the latter-day attempts, fruitless as they were, of such inveterate Cartesians as de Molières, Regnault, Castel, and Fontenelle himself, to turn back the tide of Newtonianism, were symptomatic, in pronounced form, of something that was much more widespread and effectual. Their efforts appealed to a general acceptance enjoyed by Descartes's conception of science, which thereby was on successive occasions brought more explicitly to the philosophes' attention. If in a moribund condition the vortex system, incessantly reworked to satisfy new observations and more exact calculations, survived after a fashion until the 1770's,[6] this must not be attributed to any special success it had in explaining physical phenomena. The longevity of the *tourbillon* hypothesis, rather than the result of any merit of its own, was owing to its greater degree of conformity, when compared to attractionist theories, with the abstract ideal of science recommended by Descartes. Concerning the obstinate loyalty to Descartes on that score, Voltaire was prompted to say: "Certain persons reject all of his doctrines in detail, and

[4] The earlier phase of the conflict between Cartesians and Newtonians is treated, in its technical details, in Pierre Brunet, *l'Introduction des théories de Newton en France au XVIIIe siècle: Avant 1738*, Paris, 1931.

[5] D'Alembert, "Discours préliminaire," *Œuvres*, I, 73.

[6] In 1763, Aimé-Henri Paulian tried to salvage some remnants of Descartes's physics in the *Traité de Paix entre Descartes et Newton*, Avignon. As late as 1774, Charles-Hercule de Keranflech, in the *Observations sur le cartésianisme moderne*, persevered in the thesis of his earlier *L'Hypothèse des petits tourbillons, justifiée par ses usages*, Rennes, 1761.

nevertheless call themselves Cartesians."[7] Pierre Brunet has similarly stated that, in the first half of the eighteenth century, the espousal of Descartes's physics "was much more an attitude than a conviction."[8] If the vortices were unseated from power in France during the 1730's, it must be stressed that the broader complex of Cartesian scientific attitudes, of which they had for the most part been merely an illustration, did not, by that reason or from that time, cease to exert its pressure. Descartes's methodology, taken in its basic features and divorced from the specific content of his obsolete physics, was something in which the philosophes continued almost unquestioningly to place their trust. Because of this circumstance, it will not make much difference if, in the pages following, philosophes are quoted in company with anti-philosophes in the re-creation of that Cartesian atmosphere which subtly invested the endeavors in science of eighteenth-century France. The influence of the *Discours de la méthode*, in that respect, transcended the particular issues which, in other ways, distinguished the opposing camps of "Cartesians" and "anti-Cartesians."

It is a revealing fact that, however much the philosophes repudiated Cartesian physics, they were strangely unable to forget it completely. Their thinking in questions of natural science conformed to a characteristically contrapuntal movement, regulated by the Newton *versus* Descartes theme. It was Fontenelle, perhaps, who had given the original example. His celebrated *Eloge de Newton* had incensed the London Royal Society by daring to draw a parallel between the methodologies of Newton and Descartes on a footing of equality. Fontenelle's early comparison of the two great scientists, skillfully bringing out the diametrical opposition between a deductive-rationalist and an inductive-experimentalist ideal of science, gave the essential content of the Car-

[7] Voltaire, *Œuvres*, XXIII, 75; also, XXXV, 3.
[8] Brunet, *op. cit.*, p. 9.

tesianism that was to persist concurrently with Newtonianism among his younger contemporaries in *lumières*:

"Both persons were geniuses of the first order, born to dominate other minds and to found empires. . . . But one, soaring up audaciously, wished to place himself at the source of all things, to master their first principles by means of a few clear and fundamental ideas, in order that he might then simply descend to the phenomena of nature as to so many necessary consequences. The other, either more timid or more modest, started out with the support of phenomena, so that he might mount up to their unknown principles, being resolved to admit these latter no matter what the chain of consequences might logically determine them to be. The former began with what he understood clearly in order to discover the cause of what he saw. The latter began with what he saw in order to find its cause, be it clear or obscure. The self-evident principles of the former did not always lead him to phenomena such as they actually are; the phenomena of the latter did not always lead him to sufficiently evident principles."[9]

The pitting of Newton and Descartes against each other, with varying shades of sympathy being shown for the first or the second, came to be a commonplace device with the philosophes. It is therefore not surprising that Diderot's orientation into the problem of natural science should, likewise, have been affected by the popular controversy between Newtonians and Cartesians. The *Bijoux indiscrets* of 1748, a youthful work with a few serious passages amidst licentious frivolities, drew an ingenious parallel, in the manner of Fontenelle, between the conflicting physical doctrines and methodologies of Descartes and Newton:

"Olibri, a skilful geometer and great physicist, founded the sect of the vorticosians. Circino, a skilful physicist and

[9] Fontenelle, *Œuvres*, vi, 296-297.

great geometer, was the first attractionist. Olibri and Circino both undertook to give an explanation of nature. The principles of Olibri, at first glance, have a most seductive simplicity; they give a broad interpretation of principal phenomena, but prove false in application to details. As for Circino, he seems to depart from an absurdity; but it is only the first step that is troublesome. His system is strengthened by the same minute details which ruined that of Olibri. He follows a path obscure at the beginning, but which grows progressively clearer as he advances. The path taken by Olibri, on the contrary, clear at the outset, becomes increasingly darker as it goes on. His philosophy requires less effort than intelligence. But to be a disciple of the other, much intelligence as well as effort are needed. No preparation is necessary to enter the school of Olibri: everyone possesses the key to it. The school of Circino is open only to the best geometers. Olibri's *tourbillons* are within reach of every mind. The centrifugal forces of Circino are intended only for algebrists of the first rank. There will therefore always be a hundred vorticosians for every attractionist; and one attractionist will always be worth a hundred vorticosians."[10]

Diderot's words (which confirm, incidentally, the widespread favor retained by Descartes's physics) express a quite decided preference for the method of English experimentalism, as compared to that exemplified by Cartesian tradition. But the *Bijoux indiscrets* was composed while its author, proceeding in the steps of already famous contemporaries such as Voltaire, D'Alembert, and Maupertuis, remained still under the sway of Newton's mathematical science. The speculative limitations imposed by the latter were soon after to be cast off, when Diderot's intellectual individuality found its earliest utterance in the sketch of natural philosophy offered by the *Lettre sur les aveugles*. However, in terms of

[10] Diderot, IV, 162-163.

the above differentiae set up by Diderot himself, it is permissible to ask a preliminary question or two. To what extent was his own system of nature, by virtue of a pendulum-like swing to the opposite pole, actually to derive from principles at once exhibiting "a most seductive simplicity" and designed to give "a broad interpretation of principal phenomena?" In what degree, furthermore, was the school of scientific naturalism to advertise that "everyone possessed the key to it?" Whether he later said so or no, Diderot plainly knew that a science erected on the aforementioned rationalist criteria, along with the ideological uses of its suitability for popularization, represented an eighteenth-century outcome of Cartesian thought.

To understand how Descartes's scientific procedure asserted itself at the core of French materialism, one might begin by asking what the major questions of methodology were that provoked discussion during Diderot's formative years. The leading argument brought by Cartesians against Newton's physics had been the charge that its underlying assumption—the mutual attraction of corporeal masses—was in itself an obscure and unintelligible notion, despite the mathematical evidence adduced in its support. Hence such a principle, in the Cartesian view, could not be the proper foundation for a logically certain and integrated body of knowledge. As early as 1709, Saurin had cautioned that the abandonment of clear and simple ideas in the field of mechanics, such as attractionism was considered guilty of, threatened to lead natural science back "into the night of Peripatetic philosophy, from which Heaven preserve us!"[11] The Cartesian laws of motion, based on *choc*, or mechanical collision, as universal cause, were held—in that pre-Humean period which extended actually as late as the last quarter of the century—to be immediately comprehensible and equally evident to all minds. But, argued Fontenelle, from the fact

[11] *Mémoires de l'Académie des Sciences*, 1709, p. 148.

that two bodies, A and B, "are both at rest at any given distance from each other, it does not follow in any manner that they ought to go towards—or attract—one another."[12] To resort to such inexplicable and non-mechanical tendencies in bodies was regarded as equivalent to accepting back into physics Aristotle's "occult qualities," "sympathies," "horrors," and the rest, as natural causes. The marquis d'Argens, a philosophe with marked inclinations for Cartesian doctrine, commented about Newton's law of attraction to the effect that "he did about the same thing as Aristotle, who explained the power of the magnet . . . by saying that it attracted iron because it had within itself an attractive power."[13] During most of the eighteenth century in France, the same criticism was revoiced on all sides with much earnestness and nuance. Even so assiduous a student of the *Principia* as Voltaire felt obliged to concede that the accusation of occultism made against Newton was justified. But seizing the bull somewhat rashly by the horns, he went on to contend that all causes in nature were mysterious, and, leaving the way wide open for apologetic inroads into the sphere of natural science, maintained that the first thinker who had deviated from the ancient philosophy of occult causes had corrupted the intelligence of the human race.[14] Notwithstanding all this,

[12] Fontenelle, *Théorie des tourbillons cartésiens, avec des réflexions sur l'attraction* (1752), in *Œuvres*, IX, 305.

[13] D'Argens, *Philosophie du Bon Sens*, II, 47. D'Argens sided personally with the Cartesian objection to attractionist physics: "Quant à moi, s'il m'est permis de dire mon sentiment, j'avoue que je ne puis trouver extraordinaire que bien des gens aient peine à comprendre que des attractions qui ne peuvent avoir leur principe dans l'impulsion, puisqu'on les fait régner jusques dans le vuide, aient une existence réelle & véritable. Ne seroit-il pas permis de les regarder comme les *êtres de raison* des Scholastiques?" *ibid.*, 50-51.

[14] Voltaire, *Œuvres*, XLVI, 204-205. Interestingly, although Voltaire here takes the Cartesian accusation of occultism seriously in a letter to "L. C.," he would not have done as much in print. Writing also to Maupertuis, Voltaire commented on Newton's definition of the law of attraction: "C'est bien dire nettement, bien expressément, que l'attraction est un principe qui n'est point mécanique." XXXV, 7.

it must be emphasized that the main difference between Newtonians and Cartesians had really nothing to do with the phenomenon itself of gravitation, nor with Newton's mathematical schematization of the data concerning it—both of which the defenders of *tourbillon* physics accepted as perfectly well-established.[15] The disagreement was fundamentally about what kind of *cause* ought to be assigned to explain the facts and figures at hand. The Cartesians insisted on the necessity of giving a purely mechanical and "conceivable" cause. The Newtonians were for the most part satisfied with the principle of attraction insofar as it remained simply a mathematical function, however unintelligible it might have seemed as a physical force.

The foregoing question has been discussed with a view to illustrating that Diderot did, in fact, feel the weight of the Cartesian argument. A passage in the *Premier mémoire de mathématique*, of 1748, contains his refusal to recognize in the law of attraction the cause for the ponderability of bodies.[16] But while Newton's hypothesis is there blamed for its alleged occultism, and for an unwillingness to seek out the real causes in nature, Diderot's judgment (with a skepticism akin to Hume) remains independent enough also to dismiss vortex theory as being chimerical:

"The more hidden the cause of a phenomenon, the less are the efforts made to discover it. But this laziness or discouragement of minds is neither the only, nor perhaps the greatest, obstacle to the perfection of the arts and sciences. There is a kind of vanity that prefers to attach itself to words, to occult qualities, or to some frivolous hypothesis, rather than to confess ignorance; and this vanity has even more dangerous consequences. For better or worse, one wants to explain everything. And it is thanks to this unbridled de-

[15] E.g., Fontenelle, *Œuvres*, IX, 315; Keranflech, *Hypothèse des petits tourbillons*, pp. 29, 65.

[16] Cf., Pommier, *Diderot avant Vincennes*, p. 70.

sire, that the horror of a vacuum has caused water to rise in pumps; that the vortices have been the cause of the movements of the heavens; that attraction will for a long time still be the cause of the weight of bodies. . . ."[17]

Diderot's ability to see all sides of a problem has thus led him to accredit, not only the opinion of Voltaire and other Newtonians regarding the futility of the *tourbillons*, but also the objection made by Cartesians that attractionist physics was a relapse into the use of occult qualities. While rejecting the vortices as an unworkable hypothesis, Diderot seemed nonetheless to be sufficiently influenced by Descartes's conception of science to insist, as regarded specifically Newtonian physics, on the inadequacy of giving either an "occult" or a merely mathematical account of causation. Diderot was ready to agree with many of his contemporaries that Newton had, in fact, retrogressively taken the position of Aristotle against Descartes.[18]

Another and more probing example may be given of Diderot's dissatisfaction with Newtonianism. It is known that Descartes had built his physical system on the simple and clear notion that a body in motion would, if nothing obstructed its path, continue to move *ad infinitum* in a straight line without diminution of movement. Although Newton had concurred in this presumably self-evident law, the Cartesians contended that his own principle of attraction was in effect a violation of it. For, if the attractive force acted, as Newtonians held, in a vacuum—or in empty space—and therefore without any mechanical interference, it would be expected to manifest, according to the concept of inertia just mentioned and the law of the conservation of momentum, a uniform effect entirely regardless of distance. Yet this was observed not to be the case, since the principle of attraction acted with inversely variable, rather than constant, pull de-

17 Diderot, IX, 115.

18 *ibid.*, XV, 441.

pending on the interval given between two bodies. The Cartesians claimed, therefore, that if the attractive force was modified in relation to distance, the *plenum* would have to be adopted instead of Newtonian space in order to reconcile fact with logic. Of special interest for us in this controversy is the particular conception of science that was the basis of the Cartesian *critique*. Descartes had called for a science of nature that would be at once wholly mechanical in content and completely logical in form; that is, a science in which all physical causation would be ultimately deducible from certain self-evident notions intuited, by the mind, from phenomena experienced in their most universal aspects. The extent to which Diderot understood and approved this Cartesian methodology may be judged from the following comment by him on the differences already noted between the Newtonian and Cartesian schools:

"I have told you, young man, that *qualities such as attraction are propagated to infinity when nothing limits their sphere of action.* One might object 'that I could even have said *that they would be propagated uniformly.*' To this will be added, perhaps, 'that it is hardly conceivable how a force has an effect *at a distance,* without any intermediary; that if absurdities exist at all, it is surely an absurdity to pretend that attraction in a vacuum operates differently at different distances, for nothing is perceived either within or without a body that is able to vary the effect of attraction on it. Descartes, Newton, ancient as well as modern philosophers, have all assumed that a body set in motion in a vacuum with even the smallest movement would go on to infinity uniformly in a straight line; that consequently distance in itself is neither an obstacle nor an aid; that any force whose action varies in an inverse or direct ratio with distance necessarily brings back the *plenum* and corpuscular physics; that to suppose a void and also the variability of action by any given cause results in two contradictory suppositions.'

If anyone should propose such difficulties to you, I advise you to go seek the answer from some Newtonian, for I have no idea how they are resolved."[19]

Faced with a concrete problem, Diderot thus appears definitely receptive to the requirements of Cartesian methodology, and favorable to its demand for conceiving of natural causes in terms both of mechanism and of logical deducibility. As for the role played by the *plenum* in the philosophe's approach to natural science, that will be dealt with presently.

What has so far been presented will serve as preface to the search for certain more basic affinities between Descartes's scientific method and that of Diderot. The various major themes comprising the Newtonian-Cartesian polarity, with which the thinking of Diderot and his age complied, may now, for purposes of examination, be summed up under several general headings. These are: the concept of the unity of nature; the *desideratum* of reducing all complexity of data to ultimate oneness of principle; and, lastly, the relationship between the roles of hypothesis and of observation (or experimentation) in the interpretation of nature.

Descartes's ideal of a purely deductive physics, which would originate in a certain minimum of *a priori*, self-evident notions, had the ambition of explaining nature in its entirety by means of one set of fundamental principles. His belief that science could, in effect, form a single unit composed of an unbroken chain of deductions involved necessarily the conviction that nature was itself a single, wholly interlocking entity. It resulted from the Cartesian attitude that the primary notions of any system of physics were sound and valid only if, having been conceived sufficiently broadly, they could be applied to the material world in its unitary character. By contrast with this, the method of Newtonian experimentalism was content to observe and calculate from numerous separate groups of phenomena, and

19 *ibid.*, II, 63.

to induce particular, but not always interrelated, laws in explanation of the workings of a somewhat disjointed mechanical order.

The Cartesian standpoint frequently found expression during the Enlightenment as a basis for criticism of Newtonian theories. The latter were blamed for offering at best only a fragmentary, and therefore in all probability a false—or at least undemonstrable—representation of things. One quotation taken from Keranflech's *Hypothèse des petits tourbillons* (written in defense of Cartesian physics at a moment when Newtonianism was everywhere in credit), will serve, as a sample of many similar statements, to illustrate the sense in which Descartes's standard of science survived actively in Diderot's epoch. "A system of physics is an immense affair," declared Keranflech. "It has to do with the general Plan of the Universe; with discovering the arrangement and ordering of its parts, the composition of various bodies, and the mechanism that produces all the effects one observes therein. The problem is to make all these things depend upon a single supposition, a unique principle; to find an idea whose development will be the explanation of the whole of Nature."[20] Such an affirmation of objectives remained, of course, perfectly consistent with the precedent set by Descartes himself who, at the end of the *Principes de la philosophie*, had asserted that every natural phenomenon was accounted for by his system.[21]

The desirability of a universal physics, as proclaimed by Descartes and his followers, fostered in France a strong faith in the absolute coherence of nature, whose component elements were deemed so completely integrated that no specific phenomenon was to be understood except in relation to

20 Keranflech, *Hypothèse des petits tourbillons*, p. 2.

21 *Principes de la philosophie*, iv, 199: ". . . j'ay prouvé qu'il n'y a rien *en tout ce monde visible, en tant qu'il est seulement visible ou sensible, sinon les choses que j'y ay expliquées*." Descartes, ix, 318.

the whole.[22] This tendency to view the natural order as a thoroughly interdependent unit complete in itself, and as such to make of it (regardless of whether one accepted or rejected the *tourbillons*) the point of departure for physical inquiry, had, in Diderot's thinking too, an almost axiomatic validity. In the *Entretien entre D'Alembert et Diderot*, the second of the interlocutors stated that "all things in nature are bound together, and whosoever assumes a new phenomenon or recalls a moment gone by recreates a new world."[23] The concept of the unity of nature found utterance, with ironic distortion, in the *Neveu de Rameau*. The Nephew's premises at least, if not his pessimistic inferences, speak Diderot's own mind:

"As long as one does not know everything, one knows nothing really well. It is not known where one thing goes, whence another comes, nor where both ought to be placed. One cannot tell which should go first, and which second. . . . Look, my dear philosopher, I have a feeling that physics will always be a poor science, a drop of water drawn with the point of a needle from the vast ocean, a pebble broken off the chain of the Alps! And the reasons themselves for the phenomena? In truth, it would be better to be ignorant than to know so little and so badly. . . ."[24]

The unitary vision of nature that Diderot's own scientific speculations sought forcefully to substantiate was, moreover, related in a special way to Descartes's theory of the *plenum*. Even Lucretius had been obliged, owing to his acceptance of the void, to rely on the intervention of mytho-

[22] See, for example, Keranflech, *op. cit.*, pp. 323-324.

[23] Diderot, II, 11.

[24] *ibid.*, v, 415. The Nephew's protagonist, incidentally, depreciates as ill-founded the opinion about the impossibility of physical science. At any rate, part of the *Neveu's* speech represents Diderot's own idea, as elsewhere stated by him, that, owing to the absolute interdependence among the objects of knowledge, "il semble que pour parler pertinemment d'une aiguille, il faudrait posséder la science universelle." IV, 22.

logical deities in order to give the atoms certain directions of "attraction" and "repulsion." By contrast, the *plenum* represented nature as a "closed system" in which, one hoped, every occurrence could be explained eventually in terms of the intertexture of which it was a part, without further use of outside agents. As instanced by Diderot's view of nature as "un tout," this Cartesian program offered a decisive methodological ground for eighteenth-century naturalism. In Dufour's examination of the materialist ideas rampant in the 1750's, Descartes's definition of matter, from which the *plenum* logically derived, was singled out for blame. It had resulted therefrom that material substance, "regardless of how extended it might be, forms a whole [*un tout*] whose parts, by their extremely varied motions . . . produce an admirable effect and mutually balance one another. . . ."[25] D'Argens too, discussing favorably the notion of space advocated by both Gassendi and Newton as against the *plenum*, revealed the historical antecedents of the conception that, in the Enlightenment, made of the natural order an absolutely self-sufficient and deterministic unit. D'Argens wrote: "If in reality the universe is one vast, compact mass filled up with bodies, then the slightest movement could not be made without all its corporeal parts being thereby affected. . . . It seems astonishing that when we move our finger we should agitate all the bodies extending to the outermost confines of the universe, which must necessarily be the case if the *plenum* exists and if there is no empty space."[26]

However incredible such a notion may have seemed to some persons, the postulate of nature as "un tout" proved to be a valuable asset of naturalism. Diderot himself made this clear in the *Interprétation de la nature* by applying it as a criterion to Maupertuis's *Dissertatio inauguralis*, in order to draw from that work the "terrible consequences"

[25] Dufour, *L'Ame ou le sisteme des Matérialistes*, p. 9.
[26] D'Argens, *Philosophie du Bon Sens*, II, 18; also, I, 349.

that he wished to draw.[27] In reply, Maupertuis pointed out pertinently that one of the possible meanings of "tout," in Diderot's somewhat ambiguous use of the term, had to do with the *plenum*, and with the opinion of several modern philosophers that the totality of matter made up a single *bloc*—"un continu"—without interruption among its parts.[28] The school of Leibniz, according to Maupertuis, had used thus the *plenum* to deny any possible "gap" in physical process; and, in that respect, the naturalistic implications of Descartes's initial idea presumably came down to the philosophes in part through Leibnizian thought. Be this as it may, the author of the ***Dissertatio inauguralis*** objected, against Diderot's *critique*, that the adoption of the *plenum* could not yield more than merely a "*continu* apparent." We have noted already that Diderot understood the unity of nature in the sense of an absolute continuity both causal and temporal in the world. It was specifically against this standpoint that Maupertuis, affirming the reality of Newtonian space as opposed to the *plenum's* supposed responsibility for the cosmic order, concluded that, in the universe, "it will never be necessary that something depending on the organization of one part or another will extend to the entire edifice."[29] As a concrete example, by contrast, of Diderot's position in the *plenum*-space controversy, Descartes's doctrine recurred as part of the natural philosophy outlined in the *Rêve de D'Alembert*. Mlle. de Lespinasse wanted to know why she was not equally conscious of everything occurring in herself, and in the outside world,

". . . since I am a mass of sensible points, and all things press on me as I in turn press on all things?

Dr. Bordeu: It is because impressions grow weaker depending on the distance at which they originate.

[27] Diderot, II, 48. [28] Maupertuis, *Œuvres*, II, *174.
[29] *ibid.*, *176.

Mlle de Lespinasse: If one strikes with the weakest blow at one end of a long beam, I hear this blow if I have my ear placed at the other extremity. If this beam were to touch the earth at one end and Sirius at the other, the same effect would be produced. Why is it that, with everything being related and contiguous, that is, with the beam actually existing, I do not hear what goes on in the immense space surrounding me, especially if I listen?

Dr. Bordeu: And who tells you that you do not hear it more or less? But the distance is so great, the impression so weak and so much interfered with in its transmission, and moreover you are surrounded and deafened by sounds so violent and diverse! The fact is that between Saturn and you there are only contiguous bodies, whereas there would have to be an organic continuity."[30]

The foregoing circumstances must be borne in mind to appraise properly the importance, in Diderot's scientific accomplishment, of the unitary view of nature. Curiously, Diderot has often been judged by scholars as the philosophe who was most keenly aware of disjunctive complexities in the economy of things. Guiding the search in the *Rêve de D'Alembert* for a plausible theory of transformism, for example, was the realization that "Nature has been pleased to vary the same mechanism in an infinite number of different ways."[31] Yet this was not the whole of it. Diderot's dictum, which has commonly been quoted to bear out his experimentalist—even positivist—sympathies as against any "systematic" leanings, deserves to be read in its full context. Just preceding it, the *Interprétation de la nature* plainly sets down the conviction that, beneath the endless variety exhibited by physical data—a variety that might be merely an illusory surface appearance—nature itself is essentially one. In formulating this point, Diderot expresses as an integral part

30 Diderot, II, 141-142.

31 *ibid.*, II, 15.

of his methodology the Cartesian determination to consider the object of science as an inseparable unit:

"Confusion frequently results from the fact that one supposes the existence of several prodigies where there is actually but one, because one imagines, in nature, as many particular causes as one enumerates phenomena; whereas nature has perhaps never produced but a single act. It would appear, moreover, that were nature made of necessity to produce several acts, the various results of these acts would be isolated from each other, and that universal chain of which philosophy assumes the continuity would be severed in more than one place. The absolute independence of a single fact is not compatible with the idea of the whole: and without the idea of the whole, there would be no more science."[32]

Diderot would seem to concur, here, in several methodological axioms typical of the Cartesianism prevalent in the Enlightenment. These may be stated as: (1) the belief that it was fallacious to frame (as Newtonians were repeatedly accused of doing) different principles for the elucidation of different sets of facts; (2) the insistence on the complete interdependence of physical events; and (3), the claim that the concep-

[32] *ibid.* Diderot was imbued with this view from the threshold of his intellectual career. In the translation (1745) of Shaftesbury's *An Inquiry concerning Virtue and Merit,* the philosophe was prompted by a suggestion in the text to elaborate his own thoughts in a footnote: "Dans l'univers tout est uni. Cette vérité fut un des premiers pas de la philosophie, et ce fut un pas de géant. . . . Toutes les découvertes des philosophes modernes se réunissent pour constater la même proposition." While Shaftesbury had seen primarily a moral significance in the oneness of nature, Diderot's own interest in the concept had to do mainly with its scientific utility. The philosophe was more indebted to the Cartesian tradition exemplied by the *Monde* than he cared to admit, when he went on to say: "Tous les auteurs de systêmes, sans en excepter Epicure, la supposaient [unity of nature], lorsqu'ils ont considéré le monde comme une machine, dont ils avaient à expliquer la formation, et à développer les ressorts secrets. Plus on voit loin dans la nature, et plus on y voit d'union. Il ne nous manque qu'une intelligence, et des expériences proportionnées à la multitude des parties et à la grandeur du tout, pour parvenir à la démonstration." I, 26-27.

tion of nature as *un tout* was the necessary condition and starting-point of natural science—this being, of course, essentially the converse of Descartes's doctrine of the unity of science.[33] Diderot's position in methodology, while directing attention to the extreme intricacy of observable facts, did not fail at the same time to call for the discovery of their uniform cause or generic principle. Notwithstanding the encyclopedist's intense curiosity about data as such, his main objective in the investigation of phenomena, like the Cartesians', actually took form and flesh from a unitary vision of nature.[34] Turgot said of Descartes that he had "envisaged nature like a man who, plunging a vast look down upon her, saw all of her at once."[35] Such was to be the practice of Diderot and his associates in scientific naturalism.

Descartes's proposal to explain the material world in its totality by virtue of a single group of self-evident principles incurred, moreover, on his part an assumption of unquestionable, and unquestioned, cogency: namely, that the truth is necessarily simple in essence. Since the Renaissance, in effect, the quest for simplicity in the elucidation of phenomena had received, owing largely to the revolutionary success of Co-

[33] "Il faut donc bien se convaincre que toutes les sciences sont tellement liées ensemble, qu'il est plus facile de les apprendre toutes à la fois, que d'en isoler une des autres. Si quelqu'un veut chercher sérieusement la vérité, il ne doit donc pas choisir l'étude de quelque science particulière: car elles sont toutes unies entre elles et dépendent les unes des autres. . . ." Descartes, *Regulae ad directionem ingenii*, éd. et trad. G. Le Roy, Paris, p. 7; x, 361.

[34] Venturi's *Jeunesse de Diderot*, p. 284, declares: "Le point de départ de toute sa conception reste l'unité de la nature, la croyance qu'une seule réalité se trouve à la base de tous les phénomènes." The author tends, however, to make this crucial factor in Diderot's orientation too exclusively the outcome of his "feeling" or "enthusiasm" for nature, qualifying it as "un acte de foi, un mythe . . . une nécessité à la fois logique et sentimentale." Diderot surely did bestow on an abstract philosophical concept his own emotional and temperamental coloration. But quite apart from that, his conception of nature and the scientific method it incurred both remained intimately related, in a logical as well as historical sense, to the substratum of Cartesian ideas in the eighteenth century.

[35] Turgot, "Histoire des progrès de l'esprit humain," *Œuvres*, II, 278.

pernican astronomy, an ever-widening stimulus and prominence. It was due to Descartes, however, that this typical Renaissance attitude, like several others, was raised from the status of an implicit faith, and incorporated into a consistent body of philosophical doctrine. The *Regulae ad directionem ingenii*, Descartes's most comprehensive statement of methodology, conferred on the *natura simplex* the highest degree of validity as the point of departure for scientific knowledge. Experience itself, before it could participate in the exploration of nature, was required to pass through the filter of simplicity: ". . . c'est seulement sur les choses parfaitement simples que l'expérience peut être considérée comme sûre."[36] Descartes's metaphysics had made simplicity of action a corollary to God's perfections: in the physics, purporting to spring from metaphysical roots, the obvious inference was drawn about the similar character of the laws operating in the universe. Through the influence of Cartesianism in France, the *esprit simpliste* was destined to become—quite apart from the acceptance or rejection of various portions of Descartes's metaphysics—one of the omnipresent prepossessions of the age of *lumières* in the domain (among others) of natural science.

Cartesians in the eighteenth century made abundant use of the criterion of simplicity in criticizing Newtonian physics which, in their estimation, had failed to reduce the multiplicity of things to anything like its most elementary terms. Keranflech upheld the French philosopher against Newton by arguing that "whereas matter, together with motion by impulsion alone, composed the world of Descartes, it is necessary in the Newtonian world to take into account four factors: impulsion, attraction, matter, and space."[37] The objection of another Cartesian to the Englishman's principles was equally typical of a major cross section of thought in France:

[36] Descartes, *Regulae*, p. 71; x, 394.
[37] Keranflech, *Hypothèse des petits tourbillons*, pp. 7-8.

"Although by these laws a great number of phenomena are explained, they do not suffice as an answer to that infinite variety of effects that nature exposes to our sight. The electrical virtues in certain bodies depend on a cause other than attraction . . . Newton admits it. It is necessary, moreover, to vary the general law of attraction in order to explain the contraction of muscles, and all the movements of vegetative and animal life."[38] Compared with the *Principia*, which "introduced as many particular laws as the number of different effects considered," the same writer asserted that "the simplicity of Cartesian philosophy, which with a single, identical principle accounts for that infinite variety of effects we marvel at, is doubtlessly a powerful reason in its favor, and the most unmistakable earmark of truth."[39] If the *tourbillon* theory, under the pressure of devastating arguments and faced with the proof of its own ineffectualness, was renounced relatively late, and even then with reluctance, the reason was that it had satisfied, if nothing else, the pervasive and deep-seated passion for simplicity of the French eighteenth century.

Descartes's ideal of relating the totality of physical data to a unique cause was much too prevalent an attitude not to count, also, some of the philosophes among its adherents. The statements of D'Alembert, Diderot's close friend and collaborator, are specially illuminating. In the "Preliminary Discourse" of the *Encyclopédie*, D'Alembert defined science as "the art of reducing, so far as possible, a great number of phenomena to a single one which could be considered the source of all."[40] Just as mathematical knowledge can be stripped down by analysis to a few basic propositions, D'Alembert held that "the universe, for the person able to contemplate it from a single point of reference, would be only

38 Dambésieux, *Reflexions sur la physique moderne*, pp. 14-15. Cf., also, Paulian, *Traité de Paix entre Descartes et Newton*, II, 111.

39 Dambésieux, *op. cit.*, p. 77.

40 D'Alembert, "Discours préliminaire," *Œuvres*, I, 27-28.

a unique fact and one great truth."[41] As a specific instance of where further simplification in principles was desirable, he cited the properties of the magnet—as Cartesians, too, were fond of doing—and included in the same class "a large amount of other phenomena whose interrelationship is perhaps bound up with the general system of the world."[42] In this expansive vein, D'Alembert found some good to say even of the *tourbillons*, which had "had the singular advantage of accounting for the gravitation of bodies by means of the centrifugal force of the vortex itself";[43] whereas Newton, as everyone knew, had judged it necessary to bring into the picture a tangential, as well as a centrifugal, force. If it be imagined that D'Alembert's partiality for simplification was merely the intellectual penchant or prejudice of a mathematician, one would do well to consult on the same point a foremost experimenter of the period: the abbé Nollet. The author of the *Leçons de physique expérimentale*, speaking about gravitation, admitted the difficulty of explaining certain movements in nature by the general laws of mechanical impact, or *choc*. He did not believe, however, that this of itself authorized the calling in of an additional principle, "especially since it is known that nature exhibits as much simplicity in its causes as diversity in its effects."[44]

Diderot was sufficiently influenced by the prevalence of the *esprit simpliste* to seek to extend it, in the goal of synthesis proposed by the *Interprétation de la nature*, from the sphere of rationalist to that of experimentalist science. Like many of his contemporaries, Diderot had an *a priori* faith in the ultimate simplicity of the phenomenal order, albeit his faith purported to concern itself directly with the objects of experience themselves, rather than with the clear and distinct abstractions that were the starting-point of Cartesian phys-

41 *ibid.*, 33. 42 *ibid.*, 28. 43 *ibid.*, 67.

44 Abbé Jean-Antoine Nollet, *Leçons de physique expérimentale*, Paris, 1743-1748, T. II (1743), 414.

ics. In the *Traité des systèmes*, Condillac's substitution of "fact" for "abstraction" had represented, it would seem, no radical attenuation of Descartes's standard of the *natura simplex*. Condillac had entrusted the archetypal fact with as great a degree of simplicity in its explanatory virtues as Descartes had accorded to his self-evident principles.[45] Consequently, when the unique key-fact was selected and applied to unlocking the mysteries of nature, it could reasonably be expected to acquire the attributes and perform the work of the unique set of abstractions. The philosophes might have conformed to the terminology of Condillac; but this only led them inevitably back to the methodology of Descartes. The *reductio ad simplicem* itself, which is of primary interest here, served to introduce a purely rationalist element into the core of Diderot's experimentalism. Acutely aware as he was of the many practical obstacles and dangers inherent to the simplification-process, Diderot, like D'Alembert, did not doubt that truth, as in the case of mathematics, was essentially simple: "It is not only in geometry, but in every art and science that the truth is identical with itself. The science of the entire universe is reduced to but a single fact in the divine understanding."[46]

On the one hand, the experimentalist credo seemed most often to orient Diderot's attention towards the investigation of effects rather than causes. On the other, his scientific objective, in character deductive rather than inductive, remained nonetheless to manipulate experience "until phenomena might be interrelated in such a manner that, one of them being given, all the others will follow."[47] Supporting this general deductivism, the *Interprétation de la nature* further advised that "it is necessary to work on the separation of causes, to decompose the product of their actions,

[45] ". . . dans tout systême, il y a un premier fait, un fait qui en est le commencement, et que, par cette raison, on auroit appelé *principe*." Condillac, *Œuvres philosophiques*, éd. G. Le Roy, Paris, 1947, I, 123.

[46] Diderot, II, 351.

[47] *ibid.*, 41.

and thus to reduce a very complicated phenomenon to a simple one."[48] Regarding the problem of science as it was perceived in his day, Diderot held fast with D'Alembert to the Cartesian purpose, unattainable as it might have seemed, of eventually correlating various isolated groups of facts to a single generative principle. He hoped thereby to constitute a complete system of nature:

> "As in mathematics, where investigation of all the properties of a curve discloses that they are simply the same property seen from different aspects; so in nature it will be recognized, when experimental physics has advanced farther, that all phenomena, whether of ponderability, elasticity, attraction, magnetism, or electricity, are but different forms of the same affection. But among the phenomena already known that are traceable to one of these causes, how many more intermediary phenomena must be discovered in order to establish their interrelationship, to fill in gaps, and to demonstrate their oneness? . . . There is perhaps a crucial phenomenon that would clarify, not only those which are already known, but even all those others still to be discovered in the course of time, and will unite them all together and form a system."[49]

Diderot adds that, lacking such a key-phenomenon, mere experiment and observation could go on indefinitely, but would succeed only in leading science into an ever-deepening labyrinth of disconnected data.

The themes so far discussed—interdependence of phenomena, unity of science, singleness of principle—proved to be basic features of the methodology that Diderot adopted in the elaboration of his natural philosophy. Indeed, the type of science best able to substantiate the concept, indispensable to materialist ideology, of a self-contained, creative Nature, called for a deductive procedure that could unfold the cosmic scheme of things from its initial physical causes. An inductiv-

[48] *ibid.*, 43. [49] *ibid.*, 42.

ism limiting itself scrupulously to empirical data was powerless, by comparison, to provide anything more than the image of a fragmentary order whose primary and generative principles remained mysterious, and beyond determination except possibly by natural theology. Diderot, Buffon, D'Holbach, and La Mettrie, insofar as their common purpose was the establishment of scientific naturalism, had no choice (however differently they might elsewhere have expressed themselves on the subject) but to prefer Descartes's methodology to that of Newton.

As for the roles in scientific inquiry of hypothesis and experiment, Diderot likewise inclined to the Cartesian pole of influence. Regarding those questions, the method actually exemplified by Cartesianism may be summarized conveniently as a somewhat forced combination of *a priori*, deductive reasoning, with the ordinary procedure of supposition and experimental verification—whatever the inconsistency was that resulted therefrom.[50] Descartes had, on the one hand, deduced from the definition of matter the laws of mechanics. But on the other, in order to explain the emergence of the particular patterns encountered in the real world, the *Monde ou Traité de la lumière* had applied these general laws within the framework of a specific hypothesis. According to this latter, it was assumed, without regard to the rigorous demands of deducibility, that matter was initially divided into cube-shaped particles possessing circular motion. Descartes had demanded of a hypothesis that it have two principal qualities: that it conform with experience and be able to predict phenomena; and that it remain consistent with certain pure and simple notions which the mind, both through experience and by its own intuitive grasp, had acquired concerning the realm of material objects.[51] The hypothesis as such was required to be neither a

50 Cf., Hyman Stock, *The Method of Descartes in the Natural Sciences*, esp. pp. 7-56, 88-95.

51 Descartes said of the general supposition underlying his construction of a world: "je desire que ce que j'écriray soit seulement pris pour

factually true nor a demonstrably necessary proposition. Its value lay chiefly in its ability to offer a more comprehensive and convincing explanation of phenomena than could be foreseen from the mere accumulation and classification of the particular facts that would suggest its framing. Descartes felt about the broad hypothesis supporting his own physical system, as expounded in the *Principes*, that it "ought not to be rejected until another and better one is found to explain all the phenomena of nature."[52] For Descartes hypotheses, as approximations of objective reality, were justifiable and valid if, despite their being neither factually true nor a link in the chain of deduction, they "take nothing from the truth of things, but only make all things much clearer."[53] And Descartes's own practice, based on such a definition, led the abbé de la Chappelle to say, in the *Encyclopédie* article "Hypothèse," that it had subsequently given to scientists a predilection for suppositions, even if this only too often degenerated into a taste for fictions.[54]

The vindication of hypothesis implicit in the Cartesian attitude ran head on into conflict, during the Enlightenment, with the standpoint of Newtonianism, constituting thereby still another factor in the Newton-Descartes polarity. It will serve as an index to the strength of Cartesian thought in France to note how the problem of hypothesis was treated by various philosophes. Newton, conforming closely to Baconian tradition, had of course anathematized all anticipations of

une hypothese, laquelle est peut estre fort éloignée de la verité; mais encore que cela fust, je croiray avoir beaucoup fait, si toutes les choses qui en seront déduites, sont entierement conformes aux experiences. . . ." *Principes*, iii, 44; IX, 123. At the same time, the hypothesis of the *Monde* was expected to contain nothing unintelligible and obscure, or what was not universally acknowledgeable as clear: "Or puisque nous prenons la liberté de feindre cette matiere à notre fantaisie, attribuons luy . . . une nature en laquelle il n'y ait rien du tout que chacun ne puisse connoistre aussi parfaitement qu'il est possible." XI, 33.

52 Letter to Père Mesland, dated May 1645, in Descartes, IV, 217.

53 Descartes, *Regulae*, p. 109; X, 412.

54 Diderot, D'Alembert, *et al.*, *Encyclopédie*, VIII, 417.

phenomena with the famous dictum: *hypotheses non fingo.* It had followed from this that "hypotheses, whether metaphysical or physical, whether of occult qualities or mechanical, have no place in experimental philosophy" according to which "particular propositions are inferr'd from the phenomena, and afterwards render'd general by induction."[55] Those who were inclined to take their *Principia Mathematica* quite literally, such as Voltaire, felt authorized to say of Descartes's *tourbillons*: "By the fact alone that this system is a pure hypothesis, it must be rejected."[56] Condillac imposed extremely severe conditions about the kind of hypothesis (for even Newton had not completely fulfilled his wish on this score) that the scientist might be allowed to utilize. It would have to be, stated the *Traité des systèmes*, a suppositious principle whose validity was guaranteed by a perfect induction from experimental facts, such as, for instance, the law of attraction was commonly held to be.[57] Condillac contrasted this, as one might expect, with the "abusive" type of hypothesis that had been at the heart of Descartes's natural philosophy: "What ought one to think, then, of Descartes's project when, with some cubes which he causes to move, he pretends to explain the formation of the world, the generation of bodies, and all other phenomena? If you permit the philosopher to attempt, from the depths of his study, to set matter in motion, he will dispose things as he pleases; nothing will resist him."[58] Judging from this, it is easy to guess how Condillac would have classified the speculations about the origins of the world and its contents which, in the works of de Maillet, Diderot, Buffon, and D'Holbach, went to make up the Enlightenment's scientific naturalism. Suffice it to remark here that, notwithstanding the ascendancy of Newton's

55 Newton, *The Mathematical Principles of Natural Philosophy*, "General Scholium," II, 392.
56 Voltaire, *Œuvres*, XXII, 511.
57 Condillac, *Œuvres philosophiques*, I, 195-196.
58 *ibid.*, 198.

physics in the period and the heavy influence wielded by Voltaire and Condillac in behalf of its methodological precepts, the ideal of a basically inductive science failed to dominate an important group of philosophes. Even D'Alembert, who belonged personally to the Newtonian school, considered the Cartesian explanation of the weight of bodies by means of the vortices to be "one of the most beautiful and ingenious hypotheses ever imagined by philosophy."[59] Nor was he averse to having observation "aided now and then by conjectures, when these are capable of furnishing broad views."[60]

The French eighteenth century undeniably possessed the gift of framing imaginative, all-embracing hypotheses, and Diderot himself was among those who defended their scientific value. Maupertuis, for instance, had supposed in organic nature (although he did not offer it explicitly as his own theory) a process of transformism, whereby a single prototype was assumed to be the source of all the diversity of living forms—a hypothesis, for the time, of truly sweeping breadth. Diderot's appreciation of this reveals a distinctly Cartesian approach to the meaning and function of hypothesis: "Whether this philosophical conjecture be admitted . . . as true, or rejected . . . as false, no one will deny the necessity of adhering to it as a hypothesis essential to the progress both of experimental physics and of rationalist science, and for the discovery and elucidation of phenomena connected with organic nature."[61] Another example will perhaps be more to the point. Helvetius, in *De l'Homme*, had maintained, consistently with experimentalist rules, that it was mandatory always to advance behind experimentation and never to precede it. He cited Descartes as a classic instance of the violation of this procedure. Diderot's reply, while stressing the indispensability of experiential verification, clearly justified Descartes's general use of hypothesis:

59 D'Alembert, *Œuvres*, I, 67.
60 *ibid.*, 25.
61 Diderot, II, 16.

". . . but are experiments made haphazardly? Is not experimentation often preceded by a supposition, an analogy, a systematic idea that the experiment will either confirm or destroy? I pardon Descartes for having imagined his laws of motion, but what I do not excuse him is his failure to verify by experiment whether or not they were in nature such as he had supposed them to be."[62]

The link between Cartesian methodology and the philosophes' naturalism will appear firmer if, along with the words just quoted, one recalls the anecdote, in the previous chapter, that expressed Diderot's marked sympathy for Descartes's attempt to reconstruct, from the laws of matter in motion, the actual state of the universe.

Strangely, Diderot has time and again been described as the philosophe having the least tolerance for hypotheses or "systems." The main reason for such an opinion has been, most likely, the presumed Baconian inspiration behind the *Interprétation de la nature*. However, a recent examination of Bacon's influence on Diderot has disclosed that, despite certain surface similarities, the basic problems discussed in the *Interprétation* remain unrelated to the English philosopher's genuine position. They indicate, even, a profound divergence from his recommendations of experimentalist method.[63] Bacon had, as is well known, indiscriminately banned hypotheses of all kinds, allowing for the framing of generalities only insofar as they were established by a gradual ascent through particulars. During the Enlightenment, this attitude was to find its faithful echo, among others, in Voltaire and Helvetius. But in Diderot's *Interprétation*

62 *ibid.*, 349. La Mettrie spoke in a similar spirit about Descartes's natural philosophy, "à laquelle, toute hypothétique qu'elle est, l'esprit humain devra tous les progrès qu'il fera à jamais dans les expériences même, dont elle a fait sentir la nécessité." *Œuvres*, II, 115.

63 Herbert Dieckmann, "The Influence of Francis Bacon on Diderot's *Interprétation de la nature*," *Romanic Review*, XXXIV, No. 4 (1943), p. 326.

de la nature, there was no counterpart to the intransigeant inductivism of the Baconian school.[64] Significantly missing, too, from the philosophe's work are the various practices that Bacon had suggested as accessories to his plan of induction, such as "prerogative instances," "tables of presence, absence, and degree," and the like.

While Diderot realized perhaps better than any of his contemporaries the intrinsic value of plain, brute facts, and the need of acquiring as many of them as possible, he did not accede, as ardent collectors of minutiae frequently do, to a merely classificatory notion of science. In large measure, he would have agreed with the judgment of a Cartesian such as Keranflech: "One sees with disappointment that in works of experimental physics so much care, wisdom, precaution, and effort yield but a knowledge of the surface and exterior of nature."[65] For Diderot also felt—and with an unmistakable note of pessimism—that unaided experimental practice by itself, even after centuries of labor, could furnish nothing more than "several isolated and disjointed fragments of the great chain that unites all things."[66] The simple acquisition of empirical data represented for Diderot a *regressus ad infinitum* of causes and effects, from which physics would have to free itself in order to give a coherent picture of the external world. It was precisely with this awareness that a deductive conception of science, reposing on the possibility of intuitively grasping the essential order of phenomena, came into Diderot's thinking. The *Interprétation de la nature* had occasion, at this stage of the argument, to distinguish between the mere observer of nature, dependent entirely on the testimony of his senses, and its true "interpreter":

". . . one of the principal differences between the observer

[64] *ibid.*, pp. 319ff.
[65] Keranflech, *Hypothèse des petits tourbillons*, p. 320.
[66] Diderot, II, 12.

of nature, and its interpreter, is that the latter starts out from the point where sense-perception and instruments abandon the former. He conjectures, on the basis of what is, what ought to be. From the order of things, he draws abstract general conclusions which possess for him all the evidence of concrete and particular truths; he rises to the essence itself of that order. . . ."[67]

It would be difficult to deny that in Diderot's statement the ideas of hypothesis and system have, in effect, merged to form a single concept, recalling by their almost synonymous status the special character ascribed by the age of *lumières* to Descartes's physics.[68]

The term "system" itself was employed by the eighteenth century, like several other "catch-all" words, with perplexing frequency, but rarely with any very definite meaning. It was mainly to clarify this confused situation that Condillac composed his *Traité des systèmes*. We could not, at the moment, do better than to consult his analysis of the problem. Condillac distinguished three types of system: first, that based on general abstract principles for which evidence was claimed *a priori* (such as the metaphysical systems of Descartes, Leibniz, Malebranche, etc.); secondly, that which followed from a supposition or conjecture and was therefore abstract in form, but which experience of particulars was expected eventually to confirm (such as Descartes's physics); and lastly, that derived purely from facts, or the inductive system of Newtonian experimentalism.[69] In the *Encyclopédie* article "Système," D'Alembert passively re-

67 *ibid.*, 53.

68 Contrary to this, Venturi (*Jeunesse de Diderot*, p. 291) remarks that much of Diderot's originality consisted in his differentiating between a "system" and a "hypothesis." But Venturi is obliged to admit, destroying his own contention, that Diderot's distinction of the two was not made explicit in the *Interprétation de la nature*, and remained something "without clarity or intellectual certainty."

69 Condillac, *Œuvres philosophiques*, I, 121-124.

produced Condillac's schematization, and, like him, declared only the third type to be a valid, or scientifically acceptable, system. He seized the occasion to remark that "Cartesianism . . . had put the taste for systems [using the term with the first and second connotations] very much in vogue"; but that "today, thanks to Newton, it seems that one has gotten over that prejudice."[70] Perhaps D'Alembert was being somewhat hasty in his judgment. The truth is that he was not wholly in agreement with Condillac, and elsewhere expressed his more personal attitude in an arresting epigram: "le principal mérite du physicien seroit, à proprement parler, d'avoir l'esprit de système, et de n'en faire jamais."[71] D'Alembert was himself one of the extremely rare philosophes who at the time exemplified this advice.[72] As for the "systems of nature" of Diderot, Buffon, Maupertuis, D'Holbach—those who both possessed and put to use the *esprit de système*—we should be inclined to class them, according to the hierarchy set up by Condillac, more or less in the second category.

Diderot made, indeed, a proper allowance for the activity of the *esprit de système*: "Have a system, I quite agree; but do not let yourself be ruled by it."[73] And elsewhere: "When one has formed in one's mind a system that has to be verified by experience, one must neither cling to it obstinately, nor abandon it too lightly. It is often imagined that one's suppositions are false, because adequate measures have not been taken to establish them as true."[74] Buffon said of the human intellect, in language that would not seem misplaced in a text by Descartes, that: "if it imitates the progress and works of nature; if it raises itself by contemplation to the most sublime truths; if it joins these together and makes a

[70] Diderot, D'Alembert, *et al.*, *Encyclopédie*, xv, 778.

[71] D'Alembert, *Œuvres*, I, 77.

[72] Although D'Alembert was sympathetic to the general enterprise of the *Monde*, and to the methodology with which it was linked (*supra*, chap. II, ftn. 149), his own *Sur le système du monde* confined itself rigorously to a physico-mathematical description of phenomena.

[73] Diderot, II, 23.

[74] *ibid.*, 40.

chain thereof; if it forms a whole, a system, by means of reflection, it will establish immortal monuments on unshakeable foundations."[75] The *Histoire naturelle, générale et particulière*, seeking to embrace and unify the spheres of cosmogony, geology, and biology with a consistent set of mechanistic formulae, was ostensibly its author's own execution of the scientific scheme he had in mind. Regarding specifically the doctrine of vitalistic transformism elaborated in the *Rêve de D'Alembert*, one may take Diderot's own word for it that the speculations put in the mouth of the dreaming D'Alembert comprised the type of system which, like Descartes's physics, left to unborn generations of researchers and experimenters the onerous task of substantiating its hypothetical insights. Dr. Bordeu remarked about the teaching in the *Rêve de D'Alembert*: "There you have a most profound philosophy. Though systematic at the moment, I believe that the more human knowledge progresses, the more it will be averred."[76]

The system that Diderot, like Buffon, recommended and essayed belongs, then, to Condillac's "class two." Although its validity did not lay, as in the first class, above experience and beyond the boundaries of nature, such a system was founded, nonetheless, on the kind of hypothesis rejected by Newton, Voltaire, and Condillac. Its chief source of inspiration, and standard model, in the eighteenth century was recognized, by general consent, to be Descartes's science taken separately from his metaphysics. Newtonians had ruled out all hypotheses except those which, representing a direct and exact induction from experimental data on a given problem, could forthwith be tested and either proved or disproved. The hypotheses buttressing the philosophes materialism were of another sort. Like those peculiar to Cartesianism, they were primarily imaginative, *a priori* constructions sug-

[75] Buffon, "Discours sur le style," in *Œuvres complètes*, éd. Flourens, Paris, 1853-1855, xii, 327.

[76] Diderot, ii, 180.

gested by certain "key phenomena" generalized into physical principles. While subject to a check against experience, their value did not hinge precisely on how "true" or "false" they might seem; they were not, in effect, expected to lend themselves readily to experimental verification. Such hypotheses were intended, above all, to initiate exploration into more or less unmapped areas of inquiry, and were "true" proportionately to their effectiveness in turning up a wealth of fresh facts, problems, conjectures, or experiments.[77] The philosophes were divided, concerning the evaluation of hypothesis so understood, into two main camps. As spokesman for the Newtonian side, Voltaire's position is highly instructive by virtue of its simultaneous hostility to materialism, and its incomprehension of the scientific use of hypothetical constructs. Commenting on Diderot's *Pensées philosophiques*, Voltaire said of the materialist's basic argument, as quoted in that work: "Motion essential to matter is only a supposition. And it is not permissible to found a system on a supposition about which it is not possible to have proof."[78]

Diderot's theories of transformism and "spontaneous generation," or his notion of "sensibilité physique" based on a distinction between *matière morte* and *matière vivante*; La Mettrie's automatist conception of organic and mental processes; Buffon's "molécules organiques" and "choc d'un

[77] The *Encyclopédie* article "Hypothèse," while aware of much abuse of Descartes's type of hypothesis, sought nevertheless to vindicate its legitimate use against the disparagements of Newton and his followers: "Il y a des vérités inconnues, comme des pays, dont on ne peut trouver la bonne route qu'après avoir essayé de toutes les autres: ainsi, il faut que quelques-uns courent risque de s'égarer, pour montrer le bon chemin aux autres." The author seemed to be justifying the contemporary speculations of Diderot, Buffon, and La Mettrie, no less than Descartes's *Monde*, when he said of such hypotheses that "en distinguant entre leur bon & leur mauvais usage, on évite d'un côté les fictions & de l'autre on n'ôte point aux sciences une méthode très-nécessaire à l'art d'inventer, & qui est la seule qu'on puisse employer dans les recherches difficiles, qui demande la correction de plusieurs siecles & les travaux de plusieurs hommes, avant que d'atteindre à une certaine perfection." VIII, 417-418.

[78] Cf., Torrey, "Voltaire's Reaction to Diderot," p. 1130.

comète"—all vital ingredients of materialist ideology—were hypotheses in the above-mentioned special sense. Methodologically considered, their role was equivalent to that of vortices, cube-shaped particles, animal-machine, subtle matter, and cognate features of Cartesianism. Moreover, intimately coupled with the "hypothetical-systematist" bent of scientific naturalism was the function assigned by Diderot, in the *Interprétation de la nature*, to a certain "esprit de divination" that characterized the "genius of experimental method." Mere terminology ought not to obfuscate Diderot's meaning. The faculty of his genius is patently an extra-experiential premonition, an intuitive grasp of the reality assumed to underlie appearances.[79] The "esprit d'invention," so often made by the philosophes synonymous with Descartes's science, was thus transformed into the insights of an "esprit de divination." This latter, although differing technically from Descartes's *intuitus purus*, was nonetheless capable, in Diderot's own attempt to interpret the external world, of doing the work performed by aprioristic deductivism in the physics of his predecessor.[80]

From the Cartesian background, some philosophes had inherited a peculiar "universality" in their scientific outlook. Thomas's *Eloge de Descartes* described rather well the orig-

79 Diderot, II, 24: "cet esprit de divination par lequel on *subodore*, pour ainsi dire, des procédés inconnus, des expériences nouvelles, des résultats ignorés." Compare with this Voltaire's canons of experimental procedure: "Il faut se conduire comme les Boyle, les Galilée, les Newton; examiner, peser, calculer et mesurer, mais jamais deviner." *Œuvres*, XLVI, 202.

80 *Vice versa*, in the application of intuitive method to actual physical investigation, as distinct from the role of apriorism—*ordo docendi*—in the demonstration of knowledge already acquired, Descartes's use of intuition could be made out to closely resemble the special gifts of Diderot's "interpreter of nature." Cf., for example, J. L. Mursell, "The Function of Intuition in Descartes's Philosophy of Science," *Philosophical Review*, 1919, p. 395: "When Descartes speaks of intuition, he is dealing with the actual practice and procedure of the expert investigator. The expert will develop and possess a power of immediately perceiving the essential factors of a complex situation."

inal model of this particular orientation. Descartes was said to have considered nature, "on the one hand, in all its greatness, as forming only a single immense work; and on the other, he pursues it in its details. Living nature and the inert; objects both insensible and organized; their different classes according to size and form; the chain of successive existences, destructions, and regenerations; the diversifications and connections of things; nothing escapes him, just as nothing astonishes him."[81] From such soil grew the methodology and the special complex of theories that made up scientific naturalism. Buffon's explanations ranged in scope from the origins of the solar system, down to the generation of all organisms peopling the earth, even the microscopic. Diderot, leaving cosmogonic details to his illustrious colleague, advanced a doctrine of cosmic transformism, which he applied more particularly to the evolution of organic patterns. La Mettrie attempted to identify all biological events, as well as moral and intellectual behavior, with the mechanical laws held to operate uniformly through the universe. D'Holbach, generalizing from such scientific speculations, codified the Enlightenment's materialist philosophy, according to which all phenomena, from the falling of a stone to the most complex actions of humans, were controlled ultimately by the properties and motions of matter.

All this was obviously "worlds apart" from the mathematically regulated, teleologically interpreted, experimentally verified world of Newtonian science. Buffon, Diderot, La Mettrie had, it is true, their experimentalist affinities, which on occasion they translated into both theory and practice. But the rise of scientific naturalism in their midst must be referred, predominantly, to the opposing rationalist influence. Because certain philosophes believed axiomatically that all things in nature were fundamentally reducible to intelligible notions concerning matter in motion, their ma-

[81] Thomas, *Eloge de Descartes*, p. 15.

terialism was destined to send its roots for sustenance into the subsoil of Cartesian methodology. Their consistency in that respect with Descartes's teaching may be recapitulated under several major topics: the insistence on treating nature, in the final sense, as a single entity; the determination to reduce all complexity of data to ultimate simplicity of principle; the free use of broad, all-inclusive hypotheses as vehicles of explanation; and lastly, an almost instinctive trust in the validity of aprioristic and deductive reasoning in the interpretation of the concrete and the particular.

Among the philosophes, the line separating rationalist from experimentalist science corresponded, roughly, with the line that divided materialism from more orthodox opinions, either Christian or deistic. Nor was this pattern of ideological filiation a haphazard occurrence. Henry More's *Enchiridion Metaphysicum*, proposing to refute mechanistic rationalism and its unfortunate materialist consequences, had early pointed out that by contrast experimental philosophy, as practiced by the Royal Society, conformed with the metaphysical proofs for the existence of an immaterial order.[82] That experimentalist, as opposed to Cartesian, method harmonized with teleological requirements was made explicit, in 1729, also in Roger Cotes's preface to the English edition of Newton's *Principia*:

"Without doubt this World, so diversified with that variety of forms and motions we find in it, could arise from nothing but the perfectly free will of God directing and presiding over all. From this fountain it is that those laws, which we

[82] More, *Euchiridion Metaphysicum*, Praefatio, II: "Haec autem omnia ità intelligi velim, ut Experimentalis Philosophiae, quam quidam imperitè cum Mechanica confundunt, pretium & existimatio nullatenus minuatur; qualem nempe celeberrima *Regia Societas Londinensis* profitetur, & in quo genere multa ac praeclara edidit Artis industriaeque suae Specimina, non solùm ad communes Vitae usus, sed, quod & praesens hoc *Enchiridion* testari potest, ad Veritatis Philosophicas maximè veréque Metaphysicas eruendas, apprimè utilia. Quorum quidem utilitas ac aptitudo ad rerum immaterialium exsistentiam demonstrandam. . . ." etc.

call the laws of Nature, have flowed; in which there appear many traces indeed of the most wise contrivance, but not the least shadow of necessity. These therefore we must not seek from uncertain conjectures, but learn them from observations and experiments. He who thinks to find the true principles of physics and the laws of natural things by the force alone of his own mind, and the internal light of his reason; must . . . suppose that the World exists by necessity, and by the same necessity follows the laws proposed. . . ."[83]

Cotes's manner of seeing the problem is indicative of the kind of argument that critics of eighteenth-century naturalism were later to adopt. There was, indeed, a certain commendable consistency in the position of apologetic thinking during the Enlightenment. Just as the principal line of defense, in the domain of religion, for revelation and the mysteries of the faith was based on factual evidence as such, and on the canons of empirical truth;[84] so, in the sphere of science, anti-materialist apologetics was grounded firmly in the bed-rock of inductive method, and in the assumed primacy and finality of phenomenal data. Pluche's *Histoire du Ciel* had connected the growth of materialism in France, during the 1730's, with the typically Cartesian conception of science that was unwilling to compromise with any "mystery" in nature. It was felt, by contrast, that a physics linked more closely to experience than to man's powers of speculative reasoning would prove more accommodating to religious dogma.[85] When Pluche declared that the procedure of experimentalists and classifiers of facts, as against that of systematists striving to explain nature *à fond*, was intrinsically compatible with the Mosaic "physics," he had in mind such names as Torricelli, Galileo, Pascal, Boyle, Halley,

83 Newton, *Mathematical Principles of Natural Philosophy*, "Preface of Mr. Roger Cotes," unpaginated.

84 On this question, see R. R. Palmer, *Catholics and Unbelievers in 18th-century France*, Princeton, 1939, pp. 77ff.

85 Pluche, *Histoire du Ciel*, II, 224ff.

Réaumur, Tournefort, Jussieu, Cassini.[86] In the case of Voltaire, who represented the deistic rather than Christian standpoint, the defense of Newtonian method on the one hand, and the condemnation on the other of the doctrines of *Telliamed*, Buffon, Maupertuis, and D'Holbach, formed really the two sides of the same apologetic coin. Similarly, in contemporary *critiques* of scientific naturalism, such as de Lignac's *Lettres à un Amériquain*, the abbé Barruel's *Lettres Helviennes*, and numerous others, it was a commonplace to stress the point that Buffon, Diderot, and their colleagues had strayed far from the paths of experience, experiment, and induction. Such polemical works protested, moreover, the gratuitous character of the several hypotheses with which the philosophes in question, confirmed in that deplorable *esprit de système*, seemed bent on elucidating everything in nature by means of a few abstract principles.

The scientific orientation peculiar to Diderot and the materialist school may be illustrated, more sharply still, by referring to the actual sympathies and antipathies expressed in the *Interprétation de la nature* for then living French scientists. First, the *Interprétation* showed respect for, as well as a considerable debt to, Buffon's various theories. Contrary to his supposed Baconian counsels, Diderot singled out for lengthy discussion only one work: the *Dissertatio inauguralis metaphysica de universali naturae systemate* (published pseudonymously by Maupertuis in 1751), whose title alone would cause a genuine experimentalist to shudder. Diderot, nonetheless, admiringly described the treatise as "a bold enterprise concerning the universal system of nature, and the attempt of a great philosopher."[87] As counterpart to this, Diderot attested a striking lack of appreciation for the efforts of an investigator such as Réaumur,[88] who typi-

86 *ibid.*, 356-357.

87 Diderot, II, 49.

88 The *Lettre sur les aveugles* opened with a patronizing reference to Réaumur's observation of the first reactions of the famous *aveugle-né* whose sight, by a newly perfected technique of operating for cataracts,

fied, however, the plodding, patient observer lauded in those parts of the *Interprétation de la nature* that are most reminiscent of Bacon. There were, to be sure, certain personal motives for the antagonism between Diderot and Réaumur; but beyond that, there were deeper ideological differences. Réaumur was an outstanding example of the type of experimentalist, prized by apologists, who saw in each new observation a fresh proof for the existence of God and the operations of Providence. In brief, he stood for that special combination of Newtonian method with pious concerns which, in the eighteenth-century *milieu*, was in diametrical opposition to the aims of scientific naturalism. Nor is it surprising that Réaumur wrote many of the pages for the abbé de Lignac's compendious attack on the hypothetical portions of Buffon's *Histoire naturelle*. Réaumur and Buffon represented, at the time, two antithetical conceptions of natural science.[89] The materialists, in turn, evaluated the former's lifetime labors in suitably unflattering terms. Grimm said of the great entomologist: "he was a man of small views, who did not lack merits, but whose weak and ill-assured glance did not dare to envisage nature."[90] Réaumur's monumental work, the *Mémoires pour servir à l'histoire des insectes*, was doubtless the target of Diderot's remark to the effect that the tireless accumulator of minutiae was someone to be pitied: "It must be admitted that among these experimenters there are some quite unfortunate cases; one of them will employ his whole life observing insects and will see nothing new." Diderot's admiration went out, instead, to the kind of

had been restored. In Diderot's opinion, whatever Réaumur was capable of detecting was bound to be less instructive than the epistemological theory of a "Métaphysicien à qui les principes de la Physique, les éléments des Mathématiques, et la conformation des parties seroient familiers" (éd. Niklaus, pp. 1-2, 47), which boast the *Letter on the Blind* sought itself to substantiate.

89 Cf., Jean Torlais, *Réaumur*, Paris, 1936, pp. 240-243.

90 Grimm, Diderot, *et al., Correspondance littéraire*, III, 455-456; quoted in Torlais, *op. cit.*, p. 249.

discovery rich in speculative promise, such as Trembley's polyp and Bonnet's hermaphroditic tree-louse proved then to be.[91] The *Interprétation de la nature*, in a more than supercilious tone, placed on a subordinate level the classificatory preoccupations of the renowned Réaumur:

"Persons extraordinary by their talents ought to respect both themselves and posterity by the use they make of their time. What would posterity think of us, if we had only a complete insectology to transmit to her, or an immense history of microscopic life? Great objects suit great geniuses; small ones are appropriate for lesser minds. The latter might just as well busy themselves with such matters as do nothing at all."[92]

A critic such as Lamoignon-Malesherbes, in scoring at the time the *a priori* quality of much of Buffon's theoretical constructs, took cognizance of the wide gap separating the century's experimentalist from its rationalist faction. The reader may judge for himself the applicability of the following words to the opinion voiced above by Diderot:

"These idle speculators regard the Gessners, Bellons, and Tourneforts as inferior scientists, endowed by nature with just enough energy and intellectual activity to be able to withstand great labors. So that, if their observations are ever to prove useful, they will bear fruit only at the hands of some superior genius, or some profound metaphysician who, without leaving his study, will grasp at a glance the hidden relationships that have escaped the limited intelligence of the observer. . . ."[93]

Several other questions have a pertinent and important bearing on the continuity of scientific method from Descartes

91 Diderot, II, 18. 92 *ibid.*, 51.

93 *Observations de Lamoignon-Malesherbes sur l'Histoire naturelle de Buffon et Daubenton*, Paris, An VI (1798), I, 39-40; quoted in Venturi, *Jeunesse de Diderot*, p. 307.

to Diderot. These are, specifically, the roles assigned to *experience*, to *mathematics*, and to *imagination* in the process of interpreting nature. All three are essential factors for properly understanding the emergence of the philosophes' naturalism, and stand in a definite relationship to certain aspects of Cartesian tradition.

There was a special sense in which Descartes's methodology may be said to have remained closer to experience than that of the experimentalist school. He had meant by experience the perceptions common to all persons of the most immediate, simple, and invariable events, rather than the artificially obtained, carefully selected experience that might better be called experimentation. It was from the former kind that Descartes had assumed, as if with the evidence of a clear and distinct idea, that the ultimate mechanical cause in nature was *choc*. When, in the course of the eighteenth century, Cartesians claimed that attraction was not a cause at all, and could not even be conceived as one, the basis of their contention was in effect the most prosaic and most constant of observations, comparable to an "idée pure et simple." This was that two bodies must touch, either directly or through some medium, in order to produce a change in each other; that, consequently, if existing in a vacuum and without impingement, they could not be imagined to have a mutual action of any sort. Notwithstanding, Newton had demonstrated, with a very exact formula borne out by calculation, that physical masses did in fact attract each other without apparent reference to the laws of mechanical impact. The Cartesian answer to this seemingly cogent proof was that Newton's principle of attraction was not a *real* (i.e. existing in nature) cause. The term "real" had, in this Cartesian usage, a special designation and force: it was intended in contradistinction with "mathematical." For example, Falconnet's preface to Fontenelle's *Théorie des tourbillons cartésiens*, a work defending Cartesian physics as late as 1752 against the Newtonian, distinguished between two realms: that of mathematics, and that of things purely physical. It was argued, from such a premise, that Newtonians had succeeded in accrediting the law of attraction by means of a too exclusive reliance on mathematics. But according to the same writer, it did not follow, merely because calculation confirmed Newton's hypothesis, that attraction existed in the world of material fact, or *in reality*.[94]

If Cartesians could during the Enlightenment claim, as they actually did, that mathematics was something imposed on physical reality and did not necessarily express its essential content, then a crucial ambiguity in Descartes's conception of science has itself been laid bare by the working of history. What was, precisely, the famous universal mathematics, the much-vaunted geometric method? Did the *mathesis universalis* mean for Descartes that the use of numbers and equations was the single indispensable instrument with which the order of *res extensa* was to be explored? Or was geometry the *model*, the classic type of reasoning to which all scientific expositions, even if materially non-mathematical, had to conform in order to be considered valid? In the domain of natural science, was mathematics a universal implement, or a universal standard? The point is of some consequence, specially for pursuing the manifold influence of Cartesianism. Students of Cartesian philosophy have in general tended, in defining Descartes's scientific method, merely to equate

94 Fontenelle, *Œuvres*, IX, 147-151. In the *Théorie des tourbillons cartésiens*, Fontenelle founded his objection to Newton's law of attraction, as well as the defense of the all-but-defunct vortices, on the Cartesian non-mathematical view of physical causation: "Quand on veut exprimer algébriquement ou géométriquement des forces physiques & agissantes dans l'univers, & qui ont nécessairement par leur nature de certains rapports & sont renfermées dans certaines conditions, il ne suffit pas d'avoir bien fait un calcul dont le résultat sera infaillible, & sur lequel on sera sûr de pouvoir compter; il faut encore, pour contenter sa raison, entendre ce résultat, & savoir pourquoi il est venu tel qu'il est. . . . Ici je demande pourquoi l'attraction suit les quarrés des distances plutôt que toute autre puissance?" *ibid.*, 311-312.

physics with mathematics—as rendered by the adjectival combination "physico-mathematical"—and have left the difficulty at that. This may in part be put down to the understandable weakness of the nineteenth and twentieth centuries to construe the past in their own image, and may or may not be a correct evaluation of Descartes's thought.[95] But the strictly historical problem remains, and asks for a solution. The principal concern here is to disclose the sense in which Cartesian science made its impression on the methodological issues confronting the philosophes.

As regards that question, several salient facts may be noted. In Descartes's system the mathematical entity, by being something innate, had been denied an objective existence in nature. Arithmetic, geometry and similar disciplines were said to treat of "extremely simple and general things, without being much concerned if those things exist or do not exist in nature."[96] Of a triangle, for instance, Descartes remarked that "perhaps nowhere in the world did such a figure exist, nor had it existed except in my thinking."[97] In Cartesian science, then, it might have been expected that experience of physical realities, as apprehended through certain clear and distinct notions, would transcend, if need be, the scope of the purely mathematical intuition.

The natural philosophy sketched in the *Monde* and *Principes* almost entirely without benefit of calculation, was itself the best proof of the basic separation, in Descartes's

[95] A suggestive discussion, running counter to conventional interpretations of "universal mathematics" in Descartes's scientific method, may be found in Alan Gewirtz, "Experience and the non-Mathematical in the Cartesian Method," *Journal of the History of Ideas*, II (1941), pp. 183-210. The author's thesis that Descartes conceived of the physicist's procedure as methodologically distinct from, and far more comprehensive than, that of the mathematician, is particularly valuable for the light it throws on certain affinities, both logical and historical, between Cartesianism and the science of Diderot's group.

[96] Descartes, *Méditations métaphysiques*, I; IX, 16.

[97] *ibid.*, V: "je puis former en mon esprit une infinité d'autres figures, dont on ne peut avoir le moindre soupçon que iamais elles ne soient tombées sous les sens. . . ." IX, 51.

method, between the different procedures proper to the mathematician and to the physicist. Descartes was willing, it would seem, to employ mathematics as an instrument whenever he could; but when he could not—which was more often the case—he used it merely as a criterion for deductive reasoning in general. His laws of motion had not been derived from numerical analysis: it was hoped at most that mathematics would be able, *post facto*, to confirm both the laws and their logical consequences. Descartes's physics thus remained "geometric" in a restricted sense. It dealt ultimately with the sizes, shapes, and motions of the components of matter, and deduced one proposition from the other with a supposedly Euclidean rigor. It had the abstract form, but not necessarily the concrete content, of geometry. As compared with the rigidly physico-mathematical limits of Newtonian science, the full range of Descartes's theorizing had treated deductively of measurable, rather than inductively of measured, objects. In this manner, an equivocal notion of geometric method was started on its career. But with the entry on the scene of Newton himself, who employed geometric method strictly in the sense of positing formulae based on actual computation, the inevitable clarification came. The mathematical proofs of Newton contradicted substantially the results of Descartes's *modus geometricus*. Mathematics appeared to establish what the geometrical mind, with its roots in a universal experience of phenomena, could neither encompass, nor express as a clear and distinct idea. By this crucial event, the divergence between an essentially mathematical and an essentially non-mathematical conception of science, implicit from the first in Descartes's physical system, was projected into the epoch of *lumières*, further contributing to its Newtonian-Cartesian polarity.

All this is important as regards Diderot and the methodology underlying scientific naturalism. In the heyday of the physico-mathematical science of which Newton was the oracle, and while attractionism was being perfected through the efforts of such eminent mathematicians as D'Alembert, Maupertuis, Bernouilli and Clairaut, Diderot announced with considerable aplomb the shortcomings of the physico-mathematical method. The *Interprétation de la nature* predicted (most erroneously, to be sure) that the success of such inquiry was approaching its end and would soon be of no further use in natural philosophy. In stating his reason for this seemingly rash view, Diderot's thinking curiously paralleled that of the Cartesian school, with respect specifically to the general dichotomy made between physico-mathematical science and physics as such. The philosophe declared, as had Descartes, that the "object [*chose*] of the mathematician has no existence in nature."[98] In saying this, Diderot was of course committing himself to the standpoint of innatist psychology. Perhaps without foreseeing the results too clearly, he was irrevocably abandoning the theory of knowledge of Locke and Condillac which, in tracing *all* ideas fundamentally to sensory experience of the outside world, had provided the epistemological underpinnings of the experimental-inductivist method. Diderot was not far, therefore, from the opinion of certain Cartesian physicists in his time, when he maintained that "the region of the mathematician is an intellectual realm, where what one takes to be rigorous truths lose absolutely that advantage when they are brought down to earth."[99] The divorce of physical reality from the mathematical symbol found embodiment, too, in the methodology accompanying the doctrine of the *Rêve de D'Alembert*.[100]

For Diderot, as for Descartes, the sphere of physical facts was found to lend itself only in limited degree to a conscientiously mathematical interpretation. The practical outcome of this, for both thinkers, was that the systematizer of nature was free to dispense with mathematics when it proved useless

98 Diderot, II, 10. 99 *ibid.* 100 *ibid.*, 180.

to his general scheme of explanation. A leading Cartesian in the eighteenth century, Terrasson, was ready to admit that "the use made by Cartesians of geometry seems to be only an elementary one, by comparison with that of the Newtonians."[101] And, while he was willing to accept the fact of gravitation on the strength of Newton's inverse-square formula, this did not destroy Terrasson's faith in the hypothetical structure of Descartes's *Monde*.[102] About the place of mathematics in Cartesian physics, the *Encyclopédie's* article on "Cartésianisme" recognized, as did everyone, that the great philosopher had early discovered and illustrated the radical application of geometry to the investigation of natural phenomena: "and he believed that by starting with several simple truths he could penetrate to the most hidden ones, and could teach physics, or the formation of all bodies, as one teaches geometry."[103] The abbé Pestré's summary has here alluded to the two distinct phases of Descartes's geometric method. The first corresponded to the physico-mathematical science present, for example, in the *Dioptrique*; the second, to physics proper, or the science of natural origins as given in the *Monde* and *Principes*. Approving, however, only the physico-mathematical connotation of geometric method, the *Encyclopédie* article went on to criticize the more comprehensive ambition of Descartes's physical system: "If it is possible to proceed geometrically in Physics, this is true only in some part or another of it, and without the hope of relating the whole [*lier le tout*]. Nature is not the same thing as measures and quantitative relations." The possibility was nonetheless left open, based on the alternate meaning of geometric method, of elaborating an essentially non-mathematical science that would "relate the whole." If Diderot and the materialist school discarded mathematics and sought to penetrate the veil of na-

[101] Terrasson, *La Philosophie applicable à tous les objets de l'esprit et de la raison*, p. 193.
[102] *ibid.*, 212-214.
[103] Diderot, D'Alembert, *et al.*, *Encyclopédie*, II, 719.

ture beyond the reaches of the physico-mathematical, their motivation was akin to Descartes's. Theirs was the hope to "lier le tout" by the aid of a method that would remain formally geometrical (i.e. deductive, and concerned with quantities and motions), but not necessarily founded on numerical evidence.

The first principles from which Diderot's chain of deduction started were, furthermore, of the same type as those at the source of Cartesian physics: namely, certain simple ideas derived from the most universal and constant forms of experience. In the *Principes philosophiques sur la matière et le mouvement*, the inclusion of motion, or energy, in the essence of material substance was based on just such an experience: to wit, everyone's awareness "that in this universe, all things are in transport or *in nisu*, or in transport and *in nisu* at the same time."[104] Likewise, the postulating of sensibility as a general property of matter was borne out, in the *Entretien entre D'Alembert et Diderot*, by "an experience which is repeated before our eyes a hundred times each day,"[105] that is, the changing of matter from an inorganic to an organic state. The remainder of Diderot's panvitalist, transformist system followed, for the most part, by deduction from these premises.

The advantage of such a non-mathematical deductivism was that it could presume to interpret a multitude of phenomena which normally escaped, like sand from a stiffly clenched fist, the grasp of physico-mathematical procedure. One of the main objectives of scientific naturalism was, in effect, to represent nature in its concrete complexity and variability. Diderot and Buffon championed, in that regard, the fundamental independence of physics—i.e. cosmogony, geology, biology, and at the time, chemistry—from the criteria of calculation. But in this, too, the Enlightenment's materialism happened to concur in the Cartesian methodol-

104 Diderot, II, 65.

105 *ibid.*, 107.

ogy which, in the *Monde* and *Principes*, had deduced concepts rather than numbers. Nor had Descartes's intention, in those works, differed greatly from the philosophes'. According to Thomas's *Eloge de Descartes*, which saw its subject from an eighteenth-century point of view, Cartesian physics had had the laudable aim of explaining nature in its concrete and not-always-calculable features. Thomas asserted that Newtonian science was "idealistic"; that its mathematical laws applied only under certain absolute conditions which, however, did not actually exist in nature. By contrast, Descartes's vortex-theory, whatever its faults, was praised for having sought to account for physical events with all their diverse interferences, variations, and modifications.[106]

An objection might be made to the similarity between Diderot and Descartes on the question of anti-mathematicism. Descartes, no doubt, had left the straight and narrow path of physico-mathematical method because of his ambition to expound, not only the world's origins, but a comprehensive natural philosophy as well. Was Diderot's *critique* of the physico-mathematical, however, pronounced in the same spirit and for identical motives? Could it not have been, rather, an argument made in the best interests of a truer experimentalism, for which even Newtonian science, by reason of its reliance on numerical symbols and equations, seemed too abstract and *a priori*? Since Bacon himself had proscribed mathematics as incompatible with a thoroughgoing inductivism, did not Diderot's strictures have the same significance? The answer is in the negative. In appraising the methodological role of mathematics, Diderot and Bacon were in reality only superficially, or verbally, in agreement. Bacon had written just prior to the tremendous success of physico-mathematical science, and without an inkling, it would seem, of its future possibilities. The *Interprétation de la nature*, on the other hand, was composed at a moment when that particular

[106] Thomas, *Eloge de Descartes*, pp. 42-43.

current of inquiry seemed to have reached its apogee, with a brilliant record of accomplishment behind and with still far-from-exhausted promise. Diderot could not, therefore, have criticized the utility of mathematics for the same reasons that Bacon had. His anti-mathematicism had quite another meaning, and must be related to a different ideological influence.[107]

A parallelism of partly biographical character might be drawn, at this point, between Descartes and Diderot. The seventeenth-century philosopher, despite his pioneering contribution to the development of analytical geometry, and his strikingly successful use of it in the field particularly of optics, had lost interest in mathematics after the publication of the "Géométrie."[108] This was doubtless owing to the inadequacy of the mathematical instrument to deal with the concrete complexity of nature which Descartes, as evidenced by the early composition of the *Monde*, had wished from the start to reduce to a general system of physics. In the eighteenth century, the non-mathematical orientation of the *Monde* and *Principes* came to be commonly noted, and often deplored, by the philosophes, who regarded Descartes's triumph in physico-mathematical practice to have been his principal scientific achievement. D'Alembert said of Descartes: "Mathematics, which he seemed to hold in small esteem, nonetheless constitutes today the most solid and least disputed part of his glory."[109] Voltaire believed that Descartes's errors in physics had begun with the neglect of mathematical criteria: "Geometry was a guide which he himself had in some measure fashioned, and which would have led him with certainty in his Physics. However, he abandoned in the end this guide, and gave in to systematizing."[110]

[107] Cf., Dieckmann, "Influence of Bacon on Diderot's *Interprétation de la nature*," pp. 306-309.

[108] Cf., Gaston Milhaud, *Descartes savant*, Paris, 1921; p. 246.

[109] D'Alembert, *Œuvres*, I, 66.

[110] Voltaire, *Lettres philosophiques* (XIV), p. 75.

But while some philosophes were thus lamenting the turn that Descartes's scientific method had taken, the same event recurred in their own midst in the person of Diderot. The latter's early interest in physico-mathematical problems had found expression in the *Mémoires sur différents sujets de mathématiques* of 1748, which attested a high degree of competence, even if no great originality, in that branch of inquiry.[111] But in the last of the *Mémoires* published in 1748, there was already a strong hint that Diderot's preoccupation with physico-mathematical science had run its course: "If mathematical subjects were once most familiar to me, to question me today about Newton would be to speak to me about last year's dream."[112] It is noteworthy that Diderot's renunciation of the Newtonian scientific ideal thus coincided, chronologically, with the eruption in his *milieu* of Cartesian-inspired materialism through the cosmogonical and biological speculations of de Maillet, Buffon, and La Mettrie, and just preceded their counterpart in his own *Lettre sur les aveugles*. This broad trend to naturalistic science was non-mathematical at its core because, like the *Monde*, it found calculation to be of limited value for presenting nature as *un tout*, and inappropriate for clarifying the mechanical process behind the formation of the world and its contents.

The distinction between a mathematical, and a physical science of nature, that had developed from the example of Descartes's physics, was discussed in Buffon's essay of 1749: "De la manière d'étudier et de traiter l'histoire naturelle." This was, in all probability, the immediate source for Diderot's own viewpoint. Because mathematics had for object certain abstractions that did not exactly correspond to anything in the material realm (as Descartes already had remarked), it ensued, for Buffon, that physico-mathematical

111 Cf., L. G. Krakeur & R. L. Krueger, "The Mathematical Writings of Diderot," *Isis*, XXXIII (1941), pp. 219-232.

112 Diderot, IX, 168-169.

science remained true ultimately, not to physical realities, but to its own primary definitions. Buffon concluded: "That which is called mathematical truth is therefore reducible to an identity of ideas, and possesses no [objective] reality." But: "physical truth, on the contrary, is in no manner arbitrary, and does not depend on ourselves."[113] Concerning the principles requisite for explaining things specifically physical, Buffon relied, like Diderot, on a simple, constant, and universal experience, such as that determining the Cartesians' notion of *choc*: "The phenomena which offer themselves daily to our sight, which follow one another and are repeated without interruption and in all cases, are the foundation of our physical knowledge."[114] It was, presumably, owing to this dichotomy between mathematics and physics proper that Buffon felt justified, in the hypothetical sections of the *Histoire naturelle*, to risk an explanation of how the world, and its features both organic and inorganic, had been generated—an enterprise likened, in turn, by various contemporaries to Descartes's *Monde*. Condillac too distinguished, although with a different intent, the hypotheses of the "physicist" from those of what he termed the "astronomer," linking the former with Cartesian scientific method, and the latter with that of Newton's mathematically-demonstrable astronomy. Condillac defined the boundary between the two types of science (which coincided roughly with the boundary then separating materialism from more traditional modes of thought) by stating that, whereas the physico-mathematical hypothesis restricted itself to what calculation

113 Buffon, *Œuvres*, I, 28.

114 *ibid.*, 29. The same methodology is discerned in two works, previously discussed, belonging to the historical background of scientific naturalism. The anonymous *Recherches curieuses* laid down as "une vérité constante" the transformation of fire into heat familiar to everyone, and went on to say (p. 4): "Après avoir posé un principe aussy simple que celui la, voulant proceder à faire voir comment les corps se meuvent & comment ils se sont formés, j'ay examiné. . . ." etc. As regards this non-mathematical deductivism, the *Dissertation sur la Formation du Monde* did not differ from the *Recherches curieuses.*

could rigorously bear out, "the physicists undertake to discover by what means the universe has been formed and is preserved, and what the first principles of things are."[115] Scientific naturalism, by its primary aim of explaining evolution in nature through a penetration of its first principles, inevitably aligned itself with Cartesian tradition in the employment of specifically physical, as opposed to physico-mathematical, hypotheses. This was the true import of Diderot's anti-mathematicism in the *Interprétation de la nature.*

There remains the question of imagination. If Diderot's circle abandoned the physico-mathematical ideal, it adhered all the more faithfully to a standard of *a priori* deduction. The importance of apriorism in scientific method entails, ordinarily, a recognition of the correspondingly important role of imagination. Although the continuity, with respect to the imaginative faculty, between Descartes and eighteenth-century naturalism cannot, owing to paucity of evidence, be more than merely hinted, it would not be amiss to consider the point in passing.

The Enlightenment, in dealing with the place of imagination in scientific inquiry, frequently showed an unwillingness, or inability, to discriminate properly between the imaginative, and the imaginary. This ambiguity, in a sense as old as history itself, is of special interest here, because it serves to emphasize the serious ambivalence in the philosophes' attitudes towards Descartes—an ambivalence reflected in his general influence on them. While it is common (if not commonplace) knowledge that the philosophes evaluated much of Cartesian philosophy as plainly imaginary or chimerical, sufficient notice has not been taken of their readiness, on occasion, to admire the great philosopher's vigorous, daring imagination, and to acknowledge its part in initiating the modern era of thought.[116] In the Lockean psychology trans-

115 Condillac, *Œuvres philosophiques*, I, 198.

116 The "Preliminary Discourse" of the *Encyclopédie*, for example,

mitted to the age of *lumières* by Condillac, however, the imaginative faculty was given an inferior classification, or, what proved tantamount, was confined categorically in its creative functions to the sphere of fine arts.[117] In the same vein, Voltaire could say pontifically: "Be careful not to be seduced by the imagination, which must be restricted to poetry and banished from physics."[118] But elsewhere, describing the contributive factors behind Descartes's revolutionary advancement of Western philosophy and science, the same writer seemed to view the matter differently: "Descartes was born with a lively and powerful imagination, which made him an extraordinary person in his private life, as well as in his manner of reasoning. This imagination revealed itself even in his philosophical works, where one finds at every moment ingenious and brilliant comparisons. Nature had almost made him a poet."[119]

There was perhaps a deeper meaning in Voltaire's judgment than he realized. At any rate, Descartes had himself suggested the kinship of the poetic imagination with the scientist's intuition. From the *Olympica*, a treatise now lost of which Baillet's *Vie de Des-Cartes* gives a valuable summary, the following significant fragment has been preserved:

---

said of Descartes: "Cet homme rare . . . avoit tout ce qu'il falloit pour changer la face de la philosophie; une imagination forte, un esprit très-conséquent. . . ." D'Alembert, *Œuvres*, I, 65.

117 The *Essai sur l'origine des connaissances humaines* (I, ii, §§17ff.) accorded to the imagination an essentially representational activity, or the power to reproduce images associated with perceptions that had already taken place; Condillac, *Œuvres philosophiques*, I, 14-15. For Condillac, consequently, there was a difference merely of degree in the cognitive process between imagination and memory (I, ii, §25); I, 16. The ability of the imaginative faculty to combine, in any manner it wished, perceptions recalled in the absence of their objects was assigned, as regarded scientific truth, a simply "ornamental" value (I, ii, §§89, 91): "l'imagination est à la vérité ce qu'est la parure à une belle personne," and, "son empire finit où celui de l'analyse commence." I, 32.

118 Voltaire, *Œuvres*, XLVI, 203. In his *Encyclopédie* article on imagination, Voltaire avoided ascribing to that faculty any constructive place in scientific inquiry. VIII, 566ff.

119 Voltaire, *Lettres philosophiques* (XIV), p. 72.

"Mirum videri possit quam sententiae in scriptis poetarum magis quam philosophorum. Ratio est quod poetae per enthusiasmum et vim imaginationis scripsere: Sunt in nobis semina scientiae, ut in silice, quae per rationem a philosophis educuntur, per imaginationem a poetis excutiuntur magisque elucent."[120]

Descartes's excerpt would appear to be vindicating the importance of creative imagination for intuiting essential scientific truths, which ratiocination or analysis alone might be powerless to disclose.[121] The particular terminology employed —*per enthusiasmum et vim imaginationis*—is, moreover, strikingly similar to what scholars have ordinarily found as being best descriptive of Diderot's own orientation in natural science.[122] Contrary to Condillac, Diderot realized, as had Descartes, that an imaginative-intuitive faculty, diversely applied and manifested, was at the fountainhead both of science and of poetry, serving to unite the cognitive and the esthetic on the level of a common extra-experiential insight: "Imagination is the quality without which one is neither a poet, a philosopher, a wit, a rational being, nor a man."[123] Such an opinion obviously represented a break with the epistemology of Locke and Condillac, and indirectly with the experimentalist methodology. As stated by the *Encyclopédie* article on "Cartésianisme," the method of induction was guaranteed logically, in contrast to rationalist physics, by the assumed soundness of empiricist psychology. Since our knowledge was considered ultimately the result of abstracting from and variously combining sense-perceptions, the scientist had the obligation, in explaining the physical world from

120 Descartes, x, 217.

121 Cf., Milhaud, *Descartes savant*, p. 58.

122 Pommier, *Diderot avant Vincennes*, p. 101, has referred the philosophe's system of nature to the promptings of an "imagination métaphysique." As for Diderot's "enthusiasm," it need not have been, as Venturi suggests, an independently emotive force, but the normal accompaniment of the imagination's role in scientific method.

123 Diderot, VII, 333.

which all sensations derived, of depending always on sensory experience itself, rather than on certain clear and simple notions purporting to determine the hidden causes or sensible properties of things in an intuitive fashion.[124]

Granted this, it turned out that empiricist-inductive procedure, in subjecting natural science to the inconclusive scope of sense-experience, could only favor thereby, not naturalism, but natural theology, which was encouraged to write its own apologetic epilogues to experimentally-sifted theories. By contrast, the efforts of Diderot and Buffon to explain all phenomena in terms purely of mechanical law led them to posit their first principles, not by induction from empirical data, but by an intuitive quest of the decisive, universal factors of experience, for which there was no analogue in Newtonian method. Diderot reduced, accordingly, to the same imaginative activity both the artist's vision, and the scientist's deductive "vue d'ensemble":

"To recall a necessary series of images such as they succeed one another in nature, is reasoning on the basis of fact. To call to mind a series of images such as their sequence would necessarily be in nature when given a certain phenomenon, is either reasoning from a hypothesis, or simulating. It would mean being a scientist, or a poet, depending on the goal that was proposed. And the poet who simulates, like the scientist who reasons, are equally and in the same sense consistent or inconsistent; for being consistent, and perceiving the necessary concatenation of phenomena, are one and the same thing."[125]

It might seem, even, that much of the poetry of the eighteenth century, conspicuously absent from its verse, was diverted instead into the type of scientific speculation supporting the materialist ideology. Be this as it may, the same linkage, re-

[124] Diderot, D'Alembert, *et al., Encyclopédie,* II, 719.
[125] Diderot, VII, 384.

garding the exercise of imagination, between the scientist and the poet was stated also in La Mettrie's *L'Homme machine*: "The handsomest, greatest, and strongest imagination is . . . the most suitable faculty for both the sciences and the arts. I do not try to decide whether a greater intelligence [*esprit*] is needed to excel in the art of Aristotle and Descartes, or in that of Euripides and Sophocles . . . but it is certain that the imagination alone, differently applied, has made for their separate triumphs and immortal glory."[126]

For Diderot the power of creative imagination to intuit fundamental truths in science was closely related to his concept of the genius, who was someone richly endowed in that faculty. One eighteenth-century observer said of Descartes—and his view was shared by the philosophes—that "hearkening only to the inspiration of Genius, he had subjected the universe to the laws of his imagination."[127] In Diderot's opinion, the procedure of the scientific genius, going beyond the inductivist ideal with its somewhat myopic assembling of data, seemed to coincide with the deductive method of Cartesianism: "The genius hastens . . . the progress of science and philosophy by the most fortunate and least anticipated discoveries. He rises upwards with eagle's flight towards a luminous truth, the source of a thousand other truths, to which will subsequently come crawling the timid crowd of cautious observers."[128] Diderot's own system of nature was, presumably, just such an ingenious discernment of the

[126] La Mettrie, *Œuvres philosophiques*, III, 37. The esteem for "scientific imagination" was deep-rooted in eighteenth-century France. Even Laplace, who contributed signally towards the perfecting of Newtonian physics, defended the role of imagination against censure by the inductivist school: "Impatient de connaître la cause des phénomènes, le savant, doué d'une imagination vive, l'entrevoit souvent avant que les observations aient pu l'y conduire. Sans doute il est plus sûr de remonter des phénomènes aux causes, mais l'histoire des sciences nous montre que cette marche lente et pénible n'a pas toujours été celle des inventeurs;" quoted in Bordas-Demoulin, *Le Cartésianisme, ou la véritable rénovation des sciences*, nouv. éd., Paris, 1874; p. 568.

[127] Fabre de Charrin, *Eloge de Descartes*, Paris, 1765, p. 38.

[128] Diderot, XV, 39.

"luminous truth" behind and within the natural order. The rọle of intuition remained an integral part of this method "of genius." Moreover, the place of imagination in the thought-process itself was made such by Diderot that, among its manifestations, Bacon found himself in company with Descartes: genius "imagines rather than sees; produces rather than finds; fascinates rather than conducts. It animated Plato, Descartes, Malebranche, Bacon, and Leibniz; and according to whether the imagination was either more, or less dominant in these famous men, it gave birth to brilliant systems, or discovered great truths."[129] Thomas's *Eloge* remarked that Descartes's physics had simplified and generalized the principles underlying phenomena, "in order to give them greater fecundity and breadth, for that is the procedure peculiar both to genius and to nature."[130] For Diderot, too, the genius's scientific insights were valid because, intuitively, he placed himself at the source and vantage point of the phenomenal order itself. Both nature and genius thus submitted to the same creative laws and impulses. The Enlightenment's naturalism, by its ingenious systematizations of the activity of a creative Nature, gave concrete expression to this attitude.

The materialists' methodology, then, was one of aprioristic reasoning in which experience, in the sense of observation, experimentation, and calculation, held a secondary rank. The abbé Pluche, who saw Newtonian method as the mainstay of theological traditions, had made a suggestive comment on Descartes's non-empiricist approach and its materialist tendencies:

"If an *aveugle-né* wished, on the advice of a Cartesian, to console himself for the privation of sight with the pleasure of studying physics . . . he would find himself in the same situation that Democritus desired for the purpose of ordering the

[129] *ibid.*, 39-40. [130] Thomas, *Eloge de Descartes*, p. 65.

world with greater liberty and ease. He would be in the state in which all meditative philosophers have been, who have believed themselves all the more capable of knowing the arrangement of the universe and its parts, because they took care to keep their eyes scrupulously shut in order that they might reason freely. That person whose reason is not distracted by the confusion of the senses ought, no doubt, to go from discovery to discovery. The torch of evidence, seemingly, will unveil everything to him."[131]

Had Diderot pondered this passage and deliberately taken up the challenge it contained, he could hardly have made a better response to Pluche's criticism of rationalist physics than that in the *Lettre sur les aveugles*. Saunderson, the *aveugle-né* whose blindness paradoxically permits him to look more profoundly into the workings of nature than those able to see, is there entrusted with stating the germinal form of Diderot's natural philosophy. Saunderson's blindness symbolizes, however, two features basic to Diderot's thinking, both of them related to Cartesian precedents. The first has to do with the exclusion of teleological factors from a science of nature owing to the indifference, as Saunderson puts it, of the "spectacle of nature" for a person born blind. The second meaning of the Saunderson symbol may be said to express the priority, in the task of laying the foundations for a system of physics, of intuition and deductive method over induction from observed facts.

Democritus, Descartes, and the *aveugle-né* Saunderson—the association of names offers a perfect clue to the influence of Cartesianism on Diderot's group. Descartes had characterized at once the method and the objective of the *Traité du Monde* with the invitation: "Permit your thoughts for a short while to travel out of this world, in order that they may behold another new world that I will cause to be born in their

[131] Pluche, *Histoire du Ciel*, II, 221.

presence somewhere off in imaginary space."[132] This Cartesian proposal found its echo in the speculative limbo where Diderot's world, also, started to take shape. The latter, in Saunderson's narration of the process, served to establish a naturalistic doctrine in opposition to one of natural theology: "O philosophers! transport yourselves with me to the confines of this universe, beyond the point at which I have touched, and where you see organized beings. Move about on that new ocean, and try to find amid its irregular agitations any vestiges of the intelligent Being whose wisdom you admire here in the world."[133]

If the earliest version of Diderot's natural science was thus expounded by an *aveugle-né*, its last and definitive statement was entrusted in the *Rêve de D'Alembert* to a man asleep. The analogy between Descartes's *intuitus purus* and the "démon familier" of Diderot's gifted interpreter of nature might suggest, perhaps, something more than coincidence in the fact that the fundamental scientific insights of both thinkers were intimately related, the one psychologically and the other literarily, to the dream-medium. The dream is at once the most convenient vehicle, and the most apparent symbol of the intuitive act. It is known that on a certain night in 1619 Descartes, having gone to bed "full of enthusiasm," supposedly perceived in a dream his "scientiae mirabilis fundamenta." Likewise, in the *Rêve de D'Alembert*, after the *Entretien* with Diderot has put him in a susceptible state of intellectual ferment recalling Descartes's before the famous "songe," D'Alembert is made to communicate, by dreaming aloud, the substance of Diderot's system of nature. The general hypothesis in the *Monde* concerning the formation of things had been described by Descartes as a "fable through which, it is hoped, the truth will not fail to appear sufficiently."[134] Diderot qualified the *Rêve de D'Alembert* in

[132] Descartes, XI, 31.
[133] Diderot, *Lettre sur les aveugles*, éd. Niklaus, pp. 43-44.
[134] Descartes, XI, 31.

comparable fashion when he said that it contained "the most extravagant sort of nonsense, but at the same time the profoundest philosophy; it was a clever stroke to have put my ideas in the mouth of a dreaming man."[185] Descartes's fable and Diderot's extravagant nonsense referred both to a broad imaginative hypothesis which, although not directly induced from experience and hence without adequate supports in fact, was nonetheless expected, in the long run, to explain the facts to which it applied better than the facts could hope to explain themselves.

Diderot's "délire philosophique," which he complained that experimentalists in his day were incapable of arousing,[186] but which he himself would seem to have used so liberally in the *Rêve de D'Alembert*, is of course not to be taken literally, or even seriously, so far as actual scientific practice is concerned. Also, the comparison of Diderot's "rêve" with Descartes's "songe" ought perhaps not to be pressed too hard. Yet, the juxtaposition of the two has its value as a clue, or symptom, pointing to certain fundamental and less fantastic affinities between Diderot's methodology and that inherent to Cartesian thought. Metaphorically speaking, the flash of genius required by Diderot for framing a fecund, all-embracing hypothesis was, like Descartes's *semina scientiae ut in silice*, elicited subjectively from the gifted interpreter, as distinguished from the mere observer, of nature. In the *Rêve de D'Alembert*, Diderot's vitalistic materialism was related ultimately, not so much to the wealth of biological and physiological data which that work set forth, as to certain prior notions regarding the sensibility of matter, its essential property of motion, its powers of development by successive transformations, and to other propositions deduced therefrom. The casting of the dialogue in the form of a dream corresponded to this aprioristic element in Diderot's scientific method. What Condillac had said in the *Traité des systèmes* turned

185 Diderot, XIX, 821.

186 *ibid.*, II, 87.

out to be applicable equally to Descartes's physics, and to Diderot's exposition of naturalism: "As for suppositions . . . the imagination makes them with so much pleasure and so little difficulty! It is from such foundations that one creates, and governs, the universe! All this costs no more than a dream, and a philosopher dreams easily."[137]

In conclusion, there was a notable similarity in the historical moments at which Descartes, and later scientific naturalism, appeared. For Descartes, as for Diderot, Buffon, La Mettrie, and D'Holbach, the patient work of experimenters and technicians had already accumulated a vast store of somewhat disconnected, but highly suggestive, materials. What was needed most in either case—in 1640 for physical, and in 1740 more particularly for biological, inquiry—was an ambitious, imaginative "system," capable of coordinating existing data and, by its very generality, of spurring and orienting future investigation and discovery. Such a system, moreover, was not to be achieved without first effecting some kind of synthesis of the divergent methods of experimentalism and rationalism. In the seventeenth century, Cartesian science had effected just such a synthesis. Descartes had not been aware of any basic incompatibility between his own procedure and that recommended by Bacon. Indeed, he had several times referred approvingly to the rules laid down by Verulam, with the purpose of furthering, entirely without misgivings, the ideal of cooperation between rationalists and experimentalists.[138] It is known, also, how keen was Descartes's interest in experiments, and how outstanding his ability in performing them.[139] From the standpoint of Cartesian thought, there was no conflict between experimentalism and rationalism.

Diderot and Buffon, in the following century, felt called

137 Condillac, *Œuvres philosophiques*, I, 122.

138 Cf., Milhaud, *Descartes savant*, chap. "Descartes et Bacon," esp. pp. 213-215.

139 *ibid.*, chap. "Descartes expérimentateur," pp. 191-212.

upon to synthesize the same opposing methodologies in order, like their predecessor, to be able to formulate a coherent natural philosophy. Buffon remarked that the study of nature actually required a combination of two intellectual qualities that seemed contradictory: "the vast views of an ardent genius who takes all in at a glance, and the small attentions of a laborious instinct that attaches itself to only one point."[140] At its best, the school of scientific naturalism undertook to harmonize the two. In transposing Bacon's distinction between the "manœuvre" and the "spéculatif" to his own ideological surroundings, Diderot in reality gave it concrete application to the methodological issue implied by the Newton-Descartes polarity. The *Interprétation de la nature* may be understood, in its attitude towards that dichotomy, as an attempt to restore an earlier thread of union, since broken. Diderot's work opened, in effect, with the announcement of just such a project, inspired by a state of affairs that "seems to occupy all our philosophers, and to divide them into two groups":

"One group has many instruments and few ideas; the other has a wealth of ideas and no instruments. The interest of truth requires that those who think should deign finally to cooperate with those who experiment, in order that the speculator might at last be free from the need to move about; that the manipulator might have a purpose in the infinite pains he takes; that all our efforts might be pooled and directed concertedly towards the resistance of nature; and that, in this sort of scientific league, each might perform the role that suits him best."[141]

The natural philosophies of Diderot, La Mettrie, and Buffon, sustaining the Enlightenment's materialist phase, represented, by and large, the putting into execution of the meth-

140 Buffon, *Œuvres*, I, 1-2.
141 Diderot, II, 9.

odological synthesis proposed by the *Interprétation de la nature*. In that way, certain philosophes merged, insofar as it was then possible, the experimentalist's fervor for brute facts and close observation with the speculative breadth and rationalist scruples inherited from Cartesian science.

CHAPTER IV

FROM CARTESIAN MECHANISTIC BIOLOGY TO THE MAN-MACHINE AND EVOLUTIONARY MATERIALISM

. . . si l'on s'en tient à l'automatisme de Descartes, systême qui est encore le plus en vogue, toutes ces phrases si usitées, où l'on identifie l'homme & la bête, sentent la matière.

—Abbé Coyer, *Lettre au R. P. Berthier sur le matérialisme*, 1759.

CHAPTER IV

FROM CARTESIAN MECHANISTIC BIOLOGY TO THE MAN-MACHINE AND EVOLUTIONARY MATERIALISM

THE severance of Descartes's physics from the restraining effect of his metaphysical doctrine had inaugurated a trend in materialist thinking. In time, this had culminated in the non-finalistic cosmogonies of certain philosophes, and in their attempts to deduce the actual state of the universe from a definition of matter and its inherent mechanical properties. Concurrently with that event, the resultant dissociation of Descartes's theory of organism from dualistic considerations had a complementary outcome. It proved instrumental in bringing about, by historically traceable influences, the man-machine notion of the Enlightenment, and the closely related theme of pre-Lamarckian transformism. Diderot, with good reason, may be regarded as the central figure in this school of naturalistic biology. But the role of La Mettrie as its prophet and guiding spirit can hardly be overstated. By his premature death in 1752, La Mettrie was not to follow to the nadir of its fortunes a current of thought that remained, nonetheless, heavily obligated to him. The author of *L'Homme machine* laid down the first and most radical materialist thesis of the period. His automatist conception, put from 1748 at the philosophes' disposal, soon found application in a fuller ideological context. It was mainly through La Mettrie's efforts that Cartesian mechanistic biology, together with its consequences for moral determinism, became the basis of a consistently materialistic view of man. Also contributing to the biological phase of scientific naturalism, but in diverse ways that ought to be evaluated individually, were Buffon, Maupertuis, and D'Holbach. Each of these thinkers, carrying out certain implications of Descartes's philosophy of organism, participated in the special progression of ideas that terminated in the complex of evolutionary materialism best set forth by Diderot.

For purposes of critical review, it is preferable to treat separately the two notions essential to Descartes's doctrine of organic nature. The first of these, already referred to briefly, was the general contention that all organic phenomena owed their formation, along with the rest of the cosmos from which they did not differ substantially, to the common laws of matter in motion. The second was the much-discussed, polemical beast-machine hypothesis, by which it was claimed that all the seemingly intelligent behavior of animals could, like their purely organic functions, be explained mechanically, without in any sense implying the existence in them of a "soul" distinct from matter. This automatist theory was, in effect, related logically enough to Descartes's rebuilding of the world, in its general as well as particular patterns, from exclusively mechanical principles. However, the two conceptions did not happen always, in the course of history between 1650 and 1750, to evolve in strict conjunction with one another. They did, nevertheless, converge eventually to beget the scientific naturalism, as it pertained to the realm of biology, of Diderot and his colleagues, regaining thereby the integrated character they had originally had in the Cartesian system.

The present chapter will investigate how and why the animal-automaton concept was broadened in such a manner that, in accord with certain adumbrations present in Descartes, it was finally made to circumscribe man himself. As a result, eighteenth-century materialists attributed all human functions, including life, sensibility, the passions, and in the end intelligence itself, to the physical constitution, or, in Diderot's favorite term, to "l'organisation." The man-machine, absorbed at the same time into the theme of hypothetically fabricating a world, gave rise to a transformistic materialism with regard to the problem of organic origins. Such a transformism proved to be nothing but the inevitable biological corollary to a cosmogony that probed, unfinalistically, the emergence of things in the successive developments ascribed to mechanical process. Accordingly, our first concern will be to follow historically the stream of ideas that, stressing certain implications of Cartesian science, converted the *bête machine* into the man-machine. By virtue of that act of generalization, Descartes's dualism was replaced, in the interest of greater simplicity, by a materialistic monism that rejected the soul's distinctness from the body, and redefined it as something fundamentally implicit in organized matter as such. After this, it will be discussed how the *homme machine*, in the speculations of La Mettrie, Diderot, and others, entered into the synthesis of an evolutionary materialism.

Scholarly opinion has recognized, even if somewhat cursorily or enigmatically, the connection existing between Descartes's *bête machine* and La Mettrie's *homme machine*,[1] doubtless because La Mettrie himself had first made known this ideological debt. As preface to the present study of the question, it would be fitting to quote *in extenso* the passage from *L'Homme machine* in which its author tersely denoted the Cartesian antecedents of his radical, scandalizing thesis. La Mettrie wrote of Descartes:

"That famous philosopher, it is true, made many mistakes. . . . But after all, he understood animal nature. He was the first to have demonstrated perfectly that animals are pure

[1] Cf., among others, Karl Marx, *Die Heilige Familie*, VI, iii, in *Werke*, Berlin, 1927-1935, Abt. I, Bd. 3, pp. 301-302; Lange, *History of Materialism*, I, 243 and II, 60, 62; Bouillier, *Histoire de la philosophie cartésienne*, I, 164; P. Naville, *Paul Thiry d'Holbach et la philosophie scientifique au XVIIIe siècle*, Paris, 1943, p. 233; L. C. Rosenfield, *From Beast-Machine to Man-Machine*, New York, 1941, pp. xxv, 22, 141ff. Mrs. Rosenfield's work gives a valuable account of the animal-soul controversy in its diverse historical phases; however, her approach to the specific problem of linking Descartes and La Mettrie is not our own.

machines. Now, after a discovery of such importance and which attests so much wisdom, how could one without ingratitude not pardon him all his errors! These latter, in my estimation, are all compensated by this frank avowal. All in all, whatever he might say about the distinction of two substances, it is obvious that this is merely a trick and a writer's ruse, intended to make the theologians swallow a poison concealed behind an analogy that strikes everyone, and which they alone fail to see. For, it is just this strong analogy which compels all scientists and competent judges to confess that those proud and vain beings distinguished more by their pride than by the name of men . . . are, at bottom, only animals and perpendicularly crawling machines."[2]

It remains to be seen that La Mettrie, whose exuberant temper often led him to utter paradoxes, was in this particular instance being unusually levelheaded, to the point of voicing a truism. To extract La Mettrie's actual meaning from the words just quoted, several problems must be dealt with beforehand. What was, precisely, the sense in which he understood Descartes's automatist biology? What was, basically, the nature of the man-machine which this interpretation, aided by analogical reasoning, had presumably engendered? In what ways and to what extent was the linkage between the beast-machine and the man-machine, as affirmed by La Mettrie, borne out concretely by the evidence of a historical continuity? What was the specific scientific or philosophical motivation immediately behind his audaciously thorough, intransigeant argument for human automatism?

[2] La Mettrie, *Œuvres philosophiques*, III, 80-81. The same point was made again in *Les Animaux plus que machines*, II, 29: "Avant Descartes, aucun Philosophe n'avoit regardé les animaux comme des machines. Depuis cet homme célebre, un seul moderne des plus hardis s'est avisé de réveiller une opinion, qui sembloit condamnée à un oubli . . . pour appliquer à l'homme, sans nul détour, ce qui avoit été dit des animaux." Daring as La Mettrie might have liked to believe himself, it will appear that there was abundant precedent for what he did.

Without prior examination of these issues, La Mettrie's statement of indebtedness would remain, despite its suggestive tone, something of an enigma, rather than a key to Descartes's influence on the philosophes.

To dispel the impression, from the start, that La Mettrie's acknowledgement of Cartesian paternity for the crucial notion behind his materialist system might, simply, have been a mystification at a time when all weapons were deemed good against authoritarians, it will suffice to refer to the critics themselves of eighteenth-century naturalism. Once that ideology had become dangerously widespread, its adversaries, no longer taking Cartesian apologetics at face value, often took up in earnest La Mettrie's claim of filiation. Denesle's *Examen du matérialisme*, in 1754, concluded its analysis of Descartes's *bête machine* with the complaint: "from this system to that of the Materialists is but a step, and the induction of one from the other is entirely natural."[3] Some years later, the theologian Bergier, patiently refuting the principles codified in D'Holbach's *Système de la nature*, observed that "nothing is more ridiculous than to argue against the spirituality of the human soul from the example of animals. . . . The wisest philosophers have renounced Descartes's opinion, because it has seemed to them impossible to explain all the operations of animals by means of pure mechanism. Today some pretend to explain in that manner the behavior even of man."[4]

Our story begins with Descartes himself. The theory of the animal-automaton, which already in the seventeenth century seemed to run threateningly counter to theologically-accredited doctrines of vegetative and sensitive soul, had been proffered by its deviser with various guarantees of pious utility. The designation of beasts as pure machines purported to favor belief in the immortality of man's soul. For if animals were held to possess souls, said Descartes, the dogma of an

[3] Denesle, *Examen du matérialisme*, I, 385.
[4] Bergier, *Examen du matérialisme*, I, 262.

after-life must inevitably be weakened by having either to admit that any form of spiritual substance could be destroyed, or, avoiding this, to allow all animals down to the gnat into eternity. Moreover, in the event that beasts were without sentiment and intelligence, divine providence could be absolved of responsibility for their supposed sufferings in a world of carnivorous human beings. This aspect of Descartes's thesis, based on the postulate that beasts were perfect automata devoid of sensibility, was warmly received in certain theological circles. The *bête machine* became, in effect, a principal prop of an entire school of apologetics. It manifested itself, among other ways, in the penchant that some saintly recluses of Port Royal had for habitually abusing animals on the grounds that these were totally without feeling. Thus construed, automatism was expounded with many refinements during the Enlightenment as a general support for several Christian dogmas. But if this phase (by far the more publicized one) of the animal-machine's career has now been mentioned, it is to insist that we will not be concerned with it here.

La Mettrie's man-machine, and its counterparts in the materialism of Diderot and his colleagues, had nothing to do, properly speaking, with the idea of purely insensitive automata. It will behoove us, with that in mind, to distinguish between the overt or official, and the clandestine or unofficial mode in which Descartes's theory evolved historically. These two divergent currents depended, respectively, on whether the theological, or the scientific, advantages of automatism were stressed. The apologetic version of the *bête machine* happened to have intrinsically little value as a hypothesis for biological research. That animals were machines without the least vestige of sentiment was in itself a theorem which, as many came to see for themselves, could neither be proved nor disproved. Despite its entertaining metaphysical possibilities, treatments of the subject did not help much to clarify the actual functioning of the organism itself. The religious connotations of the animal-machine remained scientifically sterile. But in formulating his somewhat paradoxical opinion, perhaps Descartes had not intended primarily to demonstrate the soul's immortality, for on that point his philosophy (regardless of what he might personally have believed or said he believed) showed a habitual indifference or skeptical reserve. And it is, in all likelihood, equally dubious that the *bête machine* was put forth deliberately as a means (to paraphrase Milton) of "justifying the ways of God to beasts."

By contrast with its theological uses, automatism was also the theoretical basis of Descartes's proposed inquiry into physiological and biological phenomena. In this regard, he made it clear that his concept did not deny vitality, sensation, and kindred organic attributes to animals. The beast-automaton thus possessed two possible and even antithetical meanings, the one proper to the metaphysics, the other to the physics, of its originator. Each of these meanings had a set of implications and a historical development of its own; in the Enlightenment the two came eventually into diametrical opposition. Setting aside as irrelevant the problem of consistency within the Cartesian system, it may be said that the metaphysical sense of automatism, consisting in the denial of conscious feeling and thought to the animal mechanism, was chiefly a logical inference from the distinction of two substances. In the sphere of physiology, on the other hand, the *bête machine* was principally a hypothesis designed to explain, through progressively closer analysis, the complex behavior of animate beings in terms of mechanical causes. The ideological continuity from Descartes to La Mettrie is understandable only if, in the period following the divorce of Cartesian physics from the metaphysics, the animal-automaton is viewed through its author's deep interest in assuring the future progress of biology and medicine. The tendency of scholars to fix the significance of automatism in relation to dualist metaphysics rather than mechanistic physiology has, in the past, raised serious obstacles to establishing the historical connection between Descartes and La Mettrie. The resolution of the difficulty has normally taken two turns, complementing each other: first, a refusal to believe that La Mettrie genuinely regarded man as a machine, together with a minimizing of his stated debt to Descartes; and secondly, a search for the sources of *L'Homme machine* in a Gassendo-Lockean tradition of empiricism. Both these approaches to the problem have been misleading, not only by their neglect of the alternate meaning of automatism and its potent impact on the course of biological speculation, but also because they have distorted the idea of mechanism central to La Mettrie's materialist doctrine.

Whatever the status of the animal-soul might have been according to his metaphysics, it is important to note that Descartes, in observing beasts from another angle, did not seem to make of them the wholly insensate creatures fancied by his more pious followers. This aspect of his theory of organism, which was to find affinities with materialist thought, was given, not so much in the works published during Descartes's life, as in his private correspondence, in the posthumous *L'Homme*, and very likely (for reasons to be cited presently) in a *Traité des animaux* which has not come down to us. Writing to the Earl of Newcastle, Descartes relaxed the rigorous definition of the insensitive *bête machine*, to the point of making the essential difference between beast and man consist in the latter's use of language as a sign of rationality.[5] The French philosopher conceded, moreover, that he had no specifically scientific or philosophical objection to endowing the animal organism with certain properties ordinarily reserved to spiritual substance, although adding cautiously that a "religious" difficulty might ensue therefrom. Indeed,

[5] Letter dated 23 Nov. 1646, in Descartes, IV, 574

Descartes appeared to confer a limited subjectivity on the beast-machine: ". . . one might conjecture that there is some degree of thought joined to these organs, such as we experience within ourselves, albeit theirs is much less perfect. To which I have nothing to answer, unless it be that if they think as we do, they would have an immortal soul like us. . . ."[6] In an informative letter to Henry More, Descartes further elucidated this position by distinguishing between a rational, and a corporeal soul. The latter, equatable with the organism itself, was the attribute of an animal, while the former was peculiar to man alone.[7] The corporeal soul, or physical organization, was judged capable of all functions excepting only cogitation, which led Descartes once again to refer the fundamental distinction between beast and human being to the norm of rationality. But apart from that, the automatist theory did not seem intended to deprive animals altogether of either vital principle or sensation: "Velim tamen notari me loqui de cogitatione, non de vitâ, vel sensu: vitam enim nulli animali denego . . . nec denego etiam sensum, quatenus ab organo corporeo dependet."[8]

Viewed in such a light, the *bête machine* proved favorable, not so much to the claim that animals were sheer unfeeling robots, as to the hope of eventually explaining mechanically all their multiform manifestations of life, sensibility, and an inferior (or non-reflective) intelligence. What was most remarkable about Descartes's expression of the automatist idea was that it served to introduce, latently at least, dynamism and vitality into something that had traditionally been (and was still for some time to remain) a synonym for the static and the inanimate. As certain of his disciples understood it,

[6] *ibid.*, 576.

[7] Letter dated 5 Feb. 1649, in Descartes, v, 276: ". . . distinguenda esse duo diversa motuum nostrorum principia: unum scilicet planè mechanicum & corporeum, quod à solâ spirituum vi & membrorum conformatione dependet, potestque anima corporea appellari; aliud incorporeum, mentem scilicet. . . ."

[8] *ibid.*, 278.

Descartes had put life and sentiment—and potentially intelligence—into the machine simultaneously with his mechanizing of the organism. The radical possibilities of such a step were to be realized, once the substantial distinction between *res cogitans* and *res extensa* no longer seemed self-evident, in the eighteenth-century school of materialism. Concurrently with this, the naturalistic tendency of the physiological program suggested by the beast-machine hypothesis became fused, in a quite plausible way, with the broader objective of Descartes's physics. The basic assumption of the *Monde* that matter in motion, or Nature, was capable of producing mechanisms which could, in turn, perform all the actions that animals were commonly observed to perform, had made it superfluous, in Descartes's opinion, to consider the role of a non-material factor in animal faculties and conduct. Given the circumstances of how the world and its contents had taken shape, it was thought more surprising that there should be a soul in every human body, than that there should be no spiritual principle in beasts.[9] For Henry More this was extended to mean that, if matter in motion was able to engender the structure of animals, it could also easily be the cause of sensibility and intelligence in them[10]—thereby agreeing with Epicurean philosophy.

The bearing of the animal-automaton on the career of materialist thinking may, perhaps, be made clearer, if it is juxtaposed with the cognate theory of organism sketched in the *Traité de l'homme*. Descartes's method of studying man, in that work, is to reconstruct hypothetically an animated

[9] *ibid*: ". . . atque ideo maiori admiratione dignum sit, quòd mens aliqua reperiatur in unoquoque humano corpore, quàm quòd nulla sit in ullis brutis."

[10] More said of Descartes, *Enchiridion Metaphysicum*, Praefatio, §5: "Materiam qualitercunque modificatam cogitare posse negavit: Sed, quod aeque difficile videtur, si non multò difficilius, Materiam motam in exquisitissimas illas corporum animalium structuras se posse concinnare satis apertè agnovit; nec ullo solido Argumento Materiam sentire non posse usquam demonstravit, Animaeve à Corpore distinctionem realem."

statue on the model of a machine. This human machine, notwithstanding certain obvious differences in the details of configuration, is considered in essence identical with the *bête machine*. By means of assembling and describing successively the various cogs and wheels that actuate the bodily mechanism, Descartes attempts to explain the diverse activities proper to man. In the end, everything has been accounted for by this procedure, with the exception of rationality. And if Descartes's treatise fails to discuss the special contribution of the rational soul to human behavior (despite a casual promise at the start to do so), the reason might well be that in reality his automatist conception has already given a sufficiently complete definition of man, without having evinced a need for spiritual substance. The conclusion of the *Traité de l'homme* shows that Descartes himself had, a century before La Mettrie, gone a long way in applying, by analogy, the beast-machine doctrine to the investigation of man:

"I wish you to consider . . . that all the functions which I have attributed to this Machine, such as the digesting of meats, the beating of the heart and arteries, nourishment and growth, respiration, waking and sleeping; the reception of light, sounds, odors, tastes, warmth, and other similar qualities, into the exterior organs of sensation; the impression of the corresponding ideas upon a common sensorium and on the imagination; the retention or the imprint of these ideas in the Memory; the internal movements of the Appetites and the Passions; and finally, the external motions of all the members of the body . . . I wish, I say, that you would consider all these functions as following altogether naturally, in this Machine, from the disposition of its organs alone, neither more nor less than do the movements of a clock, or other automaton . . . so that it is not necessary, on their account, to conceive within it any vegetative or sensitive Soul, nor any principle of motion and life other than its blood and animal-spirits, agitated by the heat of the fire that burns continually in its heart, and is not different in nature from the heat in non-animate bodies."[11]

This passage from *L'Homme* brings us admittedly to the threshold of La Mettrie's *L'Homme machine*, which simply went on to proclaim that what Descartes had depicted was not a statue at all but a real human being, and that rational consciousness was nothing but the crowning product of its multiple mechanical operations.

Cartesian physiology had proposed, then, to explore the phenomena of sensibility, sensation, and the various states of the "soul" as so many modifications of the organism defined as a machine. It was in the practical application of this method to questions of psychology that the striking originality of the *Traité des passions de l'âme* consisted. Descartes realized that his book was the beginning of a new outlook in the scientific study of man.[12] From the standpoint of the *Traité des passions*, it fell within the jurisdiction of physiology and biology to clarify ultimately the workings of the soul—a point that the author of *L'Homme machine* was emphatically to re-echo at the same time that he attached his materialism to Cartesian precedents.[13] The meaning of au-

[11] Descartes, XI, 201-202.

[12] The *Passions de l'âme* was prefaced by its author with the remark that he was obliged "d'escrire icy en mesme façon, que si je traitois d'une matiere que jamais personne avant moy n'eust touchée." XI, 328. La Mettrie expressed in *L'Homme machine* the inevitable lesson derived from the Cartesian psycho-physiology when he wrote: "ce n'est qu' . . . en cherchant à démêler l'ame, comme au travers des organes du corps, qu'on peut, je ne dis pas découvrir avec évidence la nature même de l'homme, mais atteindre le plus grand degré de probabilité possible sur ce sujet." *Œuvres philosophiques*, III, 6-7.

[13] La Mettrie said of the "médecins qui ont été philosophes:" "Ceux-ci ont parcouru, ont éclairé le labyrinthe de l'homme; ils nous ont seuls dévoilé des ressorts cachés sous des enveloppes, qui dérobent à nos yeux tant de merveilles. Eux seuls, contemplant tranquillement notre ame, l'ont mille fois surprise, & dans sa misere, & dans sa grandeur. . . . Encore une fois, voilà les seuls physiciens qui aient droit de parler ici. Que nous diroient les autres, & surtout les théologiens?" *ibid.*, 5-6.

tomatism for La Mettrie, Diderot, and Buffon must be determined in keeping with these circumstances. As Régis among Descartes's disciples had early pointed out, the animal-machine served primarily, not to deny life and feeling to beasts, but as a hypothesis for investigating their specific actions by reference to strictly mechanical principles.[14] The crucial issue that underlay the entire animal-soul controversy proved not to be, as many who engaged in it believed, whether or not animals were purely un-sentient, non-thinking automata. The real issue was whether, if the nature and behavior of animals could be fixed in relation to the laws of matter in motion, it would not be possible, from the physiological analogies between beast and man, to assert the same about the latter. Such was for the Enlightenment the basic problem involved in Descartes's conception of the *bête machine*. It was this question that certain philosophes, encouraged by a webwork of scientific evidence, ventured at length to answer in the affirmative.

If approximately a century elapsed before Cartesian biology culminated in an overt materialism, there were two good reasons at least for this long delay. First, the authority of the Church was plainly not willing to countenance any philosophy less accommodating to its dogmas than the fairly domesticated Epicureanism of Gassendi. It stood ready to do everything in its power (which up to 1750 or so was seriously oppressive) to destroy the germs of any naturalistic trend. Secondly, two great thinkers during the interim between Descartes and the Enlightenment, Malebranche and Leibniz, applied themselves to buttressing, with temporary success, the delicate frame of metaphysical dualism against the menacing weight of Cartesian science. Both occasionalism and

[14] "Et d'autant que les bêtes peuvent faire absolument tout ce qu'elles font par la seule disposition de leurs organes, nous avons crû qu'il estoit plus à propos d'expliquer toutes leurs fonctions par la machine, que de recourir pour cet effet à une ame dont l'existence est si incertaine. . . ." Pierre Sylvain Régis, *Système de philosophie*, Paris, 1690, II, 632.

the preestablished harmony, accounting each in its way for the perfect correspondence between soul and body without granting the materialist position, retarded the drawing of certain obvious conclusions from Descartes's automatist theory. But the influence of these two systems was finally swept aside, in eighteenth-century France, by the distinct current of ideas that flowed uninterruptedly, even if often underground, from Descartes down to La Mettrie and Diderot. In fact, the promulgation by Malebranche and Leibniz of the mechanistic interpretation of nature seemed to have the effect, for some of their readers, of counteracting occasional causes, monads, and the preestablished harmony. The efforts of Malebranche and Leibniz were perhaps bound, in the end, to be futile against a naturalism whose very inception was owing to the prior liberation of Descartes's science from his metaphysics. It was the attempt to explain the union of soul and body in terms of Descartes's mechanistic physiology that, in time, transformed the *bête machine* into the *homme machine*.

Descartes's biology had incurred, from the period of its appearance, the accusation of materialist tendencies, and had actually given occasion to materialist speculations. In the "Sixth Objections" to the *Méditations métaphysiques* compiled by a group of theologians, it was foreseen that the automatist view of animals might well suggest, analogically, an equivalent attitude towards men: "If it is true that monkeys, dogs, and elephants act by such cause in all their doings, there will be some who will say that all the actions of man are also like those of machines, and will no longer admit in him either sensation or understanding, seeing that, if the feeble reason of beasts differs from man's, it is only a question of degree and does not alter the nature of things."[15]

[15] Descartes, IX, 219. Maupertuis, among the philosophes, testified to the unfavorable impression made by Descartes's *bête machine*: "Il n'établissoit peut-être un systême si paradoxe que pour plaire aux Théologiens: il arriva tout le contraire. Ils craignirent que si l'on admet-

Not long after this imprecation, Regius, one of Descartes's staunchest followers at Utrecht, resolved (although not exactly in the manner foretold) the principal difficulty inherent in dualism. He gave a consistently physiological explanation of the soul, and initiated thereby a train of scientific thought that was to result in the progressive assimilation, through common mechanical laws, of man to beast. Ignoring Descartes's metaphysical doctrine and severing it from the sphere of biology, Regius interpreted the union in man of mind and matter in a manner that made inevitable the inference of their substantial identity. He wrote: "Mens humana, quamvis sit substantia à corpore realiter distincta, est tamen, quandiu in corpore existit, organica."[16] This definition of spirituality (taken in the concrete) by reference to the physical organism was borne out, in the opinion of the Professor of Medicine at Utrecht, by a profusion of evidence concerning the perfect reciprocity, and even interaction, between the various modifications of the soul and those of the body.[17]

Descartes was understandably annoyed by his pupil's premature divulgation of the scientific problems and materials with which he was then busy, and which were partly to form,

toit un tel méchanisme pour cause de toutes les actions des bêtes, on ne pût soutenir aussi qu'il suffiroit pour celles des hommes; & que les bêtes n'ayant point d'ame, les hommes ne pussent aussi s'en passer: on cria au scandale & à l'impiété." *Œuvres*, II, 211.

16 Henricii Regii, *Fundamenta physices*, Amstelodami, 1646, p. 246.

17 "Mens nostra, corpori arctissimè in unam substantiam unita, ejus est naturae, ut à corpore satis benè disposito diversimodè affici, & varia cogitatio à variis corporeis motibus ipsi excitari; & similiter mens, per variam spirituum animalium determinationem, corpus nostrum variè afficere possit. Atque hinc est perspicuum, quare, ex vario corporis temperamento, varii in homine soleant esse mores & cogitationes, cum inde varii in corpore siant motus: & cur anima, licèt sit incorporea, corpus tamen, cui unita est, de loco in locum, per variam spirituum motorum determinationem (quam per suam essentiam immediatè in iis efficit, ut ipsa per eam immediatè in corpore intelligit & vult) variè transferre queat." *ibid.*, 248-249. Regius was consistent enough also to reject the doctrine of innate ideas. (p. 251)

some years later, the contents of the *Traité des passions*. According to the preface of the *Principes*, the imprudent views expounded by Regius were based specifically on what he had found in the rough draft of Descartes's *Traité des animaux*, of which he had somehow secured a copy.[18] This occurrence might suggest that the *bête machine*, when considered in the framework strictly of Descartes's natural science, had implications at variance with the metaphysical uses of the concept. It might also indicate that Descartes, thus faced with the materialist development given to his automatist theory by Regius, despaired of bringing the *Traité des animaux* to a convincingly orthodox or publishable state. However this be, Descartes was embarrassed by the early publicizing of the fact that his physiological doctrine could well serve, for adepts of a thoroughgoing mechanistic philosophy, to subvert his metaphysical proof of the essential distinction between extended and thinking substance. He complained to Mersenne and Huygens of the violence done to his metaphysics,[19] and composed in reply to Regius the *Notae in programma* as a corrective against the seeds of materialism inherent in the new biology. But Descartes could not have failed to recognize, privately, that the conclusions drawn by Regius, although rash for the still pious temper of the times, were not too inconsistent with his own non-theological interest in the *bête machine* (as revealed in his correspondence), nor with the method and scope of the as yet unpublished *Passions de l'âme* and *Traité de l'homme*.

[18] Descartes, IX, 19.

[19] Descartes wrote to Mersenne (5 Oct. 1646) concerning Regius: "Mais, ce qui est le pis, au lieu qu'en ce qui touche la Physique, il a suivi exactement tout ce qu'il a creu estre selon mon opinion . . . il a fait tout le contraire en ce qui touche la Metaphysique; & en 4 ou 5 endroits, ou il en parle, il prend entierement le contrepied de ce que i'ay escrit en mes Meditations." IV, 510-511. Descartes perhaps realized that a consistent development of his physics, taken independently, might well contradict and substitute for his metaphysics.

After Descartes's death, several leading exponents of Cartesianism presented his notion of organism in a manner that tended, not so much to make of beasts inanimate and insensitive machines, as to make vitality and sensation the general qualitative effects of matter in motion. Rohault agreed with Descartes in despoiling animals of a rational soul; but everything else about them commonly recognized under the headings of vegetative and sensitive soul was not affected by this decision, except in the sense that all such faculties and attributes were made to depend solely on mechanical causation. The animal soul was not so much denied as mechanized. Rohault was clarifying the biological import of the beast-machine, and extricating this from the characteristically theological fortunes of automatism, when he wrote that: "in order to establish a more distinct notion of it, one must say that the sensitive soul of beasts consists precisely in the configuration and disposition of all their parts, particularly of the blood and animal-spirits."[20] Perrault too, in his popularization of Cartesian philosophy, was inclined to treat the problem in the same general fashion. He undertook to explain sensibility in animals, together with the modes of behavior accompanying it, on the basis of a mechanically-conceived organism.[21] It was this version of the *bête machine* that led Rohault to a curiously astute prediction (imputed to an unnamed person) about what a group of "philosophes" were someday to make of Descartes's idea: "if you now hold for certain that beasts are only simple automatons, are you not afraid of having to believe in the future that men are likewise nothing but simple machines,

[20] Jacques Rohault, *Entretiens sur la philosophie*, Paris, 1671; pp. 170-171.

[21] Claude Perrault, *Essais de physique*, Paris, 1680, Vol. III, chap. "La Mechanique des animaux." La Mettrie cited this text as formally expressing the identification of soul with organic functions. *Œuvres philosophiques*, III, 75.

as Monsieur P. assures that a new generation of philosophers will in time claim?"[22]

Régis expounded the *bête machine* doctrine, like the foregoing condisciples of Descartes's physics, in conformity with its materialist implications. His statement of viewpiont rendered still more apparent the rift between the physiological, and the metaphysical connotations of automatism: "While in explaining the functions of animals we have made no mention of their souls, but on the contrary have attributed all their actions to the arrangement of their organic parts alone, to the heat of their blood and the force of their animal-spirits, it is not our design to deprive them of either life or sentiment, nor even to refuse them a soul, provided only that by the *Soul of Beasts* is understood a soul consisting in the blood, or rather which is the blood itself, in particular its most subtle elements composing the animal-spirits."[23] Régis was unable, towards the end of the seventeenth century, to ignore the palpable divergence between his own version of the beast-automaton, and that derived from Cartesian metaphysics and then already in wide use for apologetic purposes. Descartes's disciple indicates how far dualism has evolved in the direction of a naturalistic monism. Régis informs the readers of his work that they may find new definitions of *Esprit, Ame, Entendement*, etc., for which however he disclaims the credit of great originality, since these were implied by Descartes's own thought: "all I have said having to be ascribed to *Monsieur Des-Cartes*, whose Method and Principles I have followed even in those explanations which differ from his."[24]

Even a thinker like Malebranche, notwithstanding his system of occasionalist dualism, may be said to have contributed against his wish, insofar as he elaborated on the subject of mechanistic physiology, to the materialist trend

22 Rohault, *op. cit.*, p. 191.
23 Régis, *Système de philosophie*, II, 630-631.
24 *ibid.*, I, Pref.

stemming from Cartesian science. It was Malebranche who, according to La Mettrie, had had a leading part in creating the vogue of the *esprits-animaux*—that somewhat indefinite and equivocal substance which, resembling (in Descartes's words) "a wind or very subtle flame,"[25] substituted so ubiquitously and effectively in human behavior for the spiritual soul, as almost to render the latter's role superfluous. Up to the time of Bordeu, the *esprits-animaux* were to enjoy practically complete authority in the physiological domain. It was largely on their account that La Mettrie was able to say of Malebranche: "Although he admits in man two distinct substances, he explains the faculties of the soul by those of matter."[26]

It will suffice to mention one more instance, among many others, of how Descartes's theory of organism early promoted scientific inquiry along unmistakably naturalistic lines. As La Mettrie remarks, Guillaume Lamy, a professor of the Medical Faculty at Paris and a proponent of mechanistic biology, suspected Descartes with good reason of having been an adroit materialist.[27] In his *De Principiis rerum*, Lamy pointed out after a careful analysis of Cartesian doctrine that, on several fundamental questions, metaphysical dualism could not be made to square with the general method and principles of the philosopher's physiology. Granted the machine-organism and its logical consequences, it seemed that the soul's immateriality and its essential independence from the body became quite undemonstrable propositions. Descartes's science, in that regard, was said to be basically in agreement with Epicureanism.[28] In a subsequent work, Lamy enlarged with more technical discussion of details upon the materialist possibilities of Cartesian biology. The *Explication méchanique et physique des fonctions de l'âme*

25 Descartes, XI, 187.
26 La Mettrie, *Œuvres philosophiques*, I, 242.
27 *ibid.*, 240.
28 Guillaume Lamy, *De Principiis rerum*, Paris, 1669, esp. pp. 147ff.

sensitive, ou des sens, des passions et du mouvement volontaire had for object, as its title eloquently suggests, the proof that "the sensitive soul is a very subtle body always in motion."[29] From there, the treatise went on to conclude that sensation, the passions and voluntary behavior all fell under the general laws of mechanism deduced from the notion of *choc*. "There is nobody who cannot readily understand this explanation," declared the author, "which is founded on the movement that one body is able to communicate to another according to the universal law of all nature."[30] Lamy's use of *esprits-animaux*, *ébranlement des nerfs*, and similar concepts proper to Descartes's physiology, therefore brings us into the close vicinity of La Mettrie's celebrated thesis. Lamy was aware that his ideas might be regarded by some as dangerous, on the ground that what he said about the materiality of the sensitive soul could be extended to the rational soul as well; that thus the *bête machine* would be metamorphosed into the *homme machine*.[31] Such a consideration did not dissuade him, however, from fusing together by way of conclusion the notions of *esprits-animaux* and of *âme*: "Throughout this little work, I have used indiscriminately the terms 'soul' and 'esprits,' which ought not to cause any confusion, for they are the same thing. I have frequently employed the word 'animal-spirits' to signify that portion of the soul which is contained in the nerves; and the word 'soul' to designate the animal-spirits contained in the brain."[32] The fact that official permission was granted to publish the *Explication méchanique* some seventy years before the appearance of *L'Homme machine* provoked a scandal forcing its author to flee France, may be ascribed to a thin veneer of skepticism that overlay Lamy's audacious speculations, and, more likely, to the censoring of his book

[29] Lamy, *Explication méchanique et physique des fonctions de l'âme sensitive, ou des sens, des passions et du mouvement volontaire*, Paris, 1678, p. 14.

[30] *ibid.*, 15.

[31] *ibid.*, 67.

[32] *ibid.*, 141.

by the untheological-minded Faculty of Medicine itself. But such occurrences served only to make the sentinels of orthodoxy more cautious in the future, and soon drove the current of naturalism flowing from Cartesian science into clandestine channels.

Bayle's treatment of the *bête machine* problem offers an excellent illustration of this. The article "Rorarius" of the *Dictionnaire historique et critique* did its best, short of bringing itself under censure, to convey to the reader the materialist standpoint concealed in Descartes's automatism. Bayle's technique was to start out by confronting his audience with a kind of dilemma. It was argued that, by virtue of the Scholastic granting of sensitive soul to beasts and the limited intelligence that this conferred on them, there would exist a difference merely of degree, and not of kind, between the spirituality typical of animals and that peculiar to man.[33] This presentation of the issue, Cartesian enough in its main lines, having presumably disposed the reader favorably towards the idea that animals might, after all, be automata, Bayle next takes up the alternate contention, made by critics of the *bête machine*, that beasts might be endowed with a "material soul." This concept Bayle identifies, by the dialectical skill of which he was master, with Descartes's reference to "corporeal soul" as the equivalent of the organism itself, and intimates, by analogical reasoning not unlike what was to prompt the man-machine doctrine, that the role of a material soul in animals would serve only to reduce man to the same status. This part of Bayle's discussion is put slyly in the mouth of an "habile Peripatéticien":

"Once you admit that all the most admirable things which take place in the case of beasts can be produced by means of a material soul, will you not soon be led to take the step

[33] Bayle, *Dictionnaire historique et critique*, xii, 599-601, n. "E."

and say that all things which occur with respect to man can likewise be produced by means of a material soul? . . . if you state that a corporeal principle is capable of causing all that monkeys do, I will maintain against you that a corporeal principle could be the cause of all the things that stupid persons do; and provided that matter be further subtilized . . . it will be the cause of all the things that clever people do."[34]

Bayle has here, like Regius and Lamy, run in the concept of soul with that of organism. It is apparent, from his handling of the theme, that Bayle took the beast-automaton as an invitation to extend to the study of man what was held to be valid in the study of animals. Under the circumstances, even the allegation of a sensitive soul for the latter did not necessarily alter the terms of the problem—any more than it had for Perrault, Régis, or Lamy—since the notions of sensibility and mechanism did not irrevocably contradict each other in the theory of organism inherited from Descartes.

Bayle could say of the Scholastics: "They cite against Descartes the most surprising actions of animals . . . but after this they feel that they have gone too far, and have furnished arms to their adversary with which to destroy the specific differences that they desire to establish between our soul and that of beasts."[35] There follows a thick verbiage of indifferent matters suited for relaxing the censor's guard. Then Bayle comes quickly to the point with the inference of a man-machine theory, while discussing the Benedictine Père Lami's *Connaissance de soi-même*, which had tried to solve "the most embarrassing difficulty of the automatist system." Insinuating what he wishes under guise of offering criticism, Bayle says of Lami's book that it had sought to prove "that other men are not simple machines; and this is nevertheless what one seeks to infer from the opinion that beasts are

[34] *ibid.*, 602, n. "F."

[35] *ibid.*, 603.

composed of organs so well arranged, as to render them capable of doing without intelligence everything we see them do. If God could fabricate such a machine, one replies"—and we may safely assume that "one" is none other than Bayle himself—"He could also construct other machines that would perform all the actions of man; consequently we could be sure only of our own individual thoughts, and would have to doubt if other human beings think."[36] The article "Rorarius" furnished, along with such carefully worded suggestions of the man-machine idea, a refutation of Leibniz's solution of the body-soul correspondence by the preestablished harmony.[37] In answer to this, Leibniz significantly admitted the consistency of Bayle's reasoning with Cartesian premises. He maintained, however, in justification of his own metaphysical position that the preestablished harmony was the surest way out of the difficulty proposed by the materialistic version of automatism. Leibniz recognized that certain persons had, by a manipulation of the notion of material soul, actually "extended to man what the Cartesians grant in regard to all other animals."[38] It was said, moreover, concerning the union of mind and matter that "the Cartesians had made out very poorly, more or less like Epicurus with his declination of atoms." Like Bayle, the German philosopher commented: "their manner of proving that beasts are merely automatons goes so far as to justify anyone who would say that all other men, excepting himself, are also simple automatons."[39]

Of still greater interest are the observations on Descartes's theory of organism that Bayle inserted in his article "Dicéarque." A student of Aristotle's, Dicearchus had claimed

[36] *ibid.*, 616, n. "K."

[37] *ibid.*, 608-610, n. "H"; 616-622, n. "L."

[38] These matters were made widely available to the French eighteenth century in Desmaizeaux, *Recueil de diverses pieces, par Mrs. Leibniz, Clarke, Newton.* See, "Replique de M. Leibniz aux Reflexions de M. Bayle," II, 399.

[39] *ibid.*, 400.

that the soul was identical with the corporeal organization. He had even taught, in a fashion reminiscent of Diderot's concept of *sensibilité*, that the principle of life was inherent to all bodies, regardless of the relative state of animation or inanimateness in which they happened momentarily to be. Bayle took this opportunity to draw a parallel between the Dicearchean panvitalism and Descartes's philosophy of organism. The analogy between the two is made on the pretext of reporting an "objection" voiced by an anonymous "Philosophe" who, no doubt, is a figment of Bayle's prudence:

"He pretends that Dicearchus meant to say that living bodies differ from the non-living only in that their parts are figured and arranged in a certain manner. He compares this opinion with that of Descartes, and here is how. If a dog is different from a stone, the reason for this is not that the former is composed of a soul and a body, while the latter is merely a body. The one and only reason is that the dog is composed of parts arranged in such a way that these constitute a machine, which is not the case for the disposition of the corpuscles in a stone. That is the belief of Descartes. Such an idea is very suitable for enabling us to understand the doctrine of Dicearchus. We have only to suppose that he extended to every kind of living body what the Cartesians assert with respect to beasts alone. We need merely assume that he reduced man to the condition of a machine, whence it would result that the human soul is not at all distinct from the body, but is simply a machine-like construction or arrangement of the various particles of matter."[40]

This passage is of capital importance for ascertaining the role played by Descartes's biology in the genesis of the philosophes' naturalism. Two concepts central to Cartesian thought are here seen to converge by their refraction through Bayle's mind. First, the *bête machine* is made the basis for

[40] Bayle, *op. cit.*, v, 512, n. "L."

attributing the soul's manifestations to mechanical modifications in the body, thus tending to define the one in relation to the other. Secondly, Descartes's distinction between the organic and the inorganic, consisting as it did solely in the ordering of corporeal elements, is made to convey a panvitalist view of nature. Vitality and sensibility are, in consequence, conceived as dependent upon the combinations of moving matter which, according to Cartesian physics, went to make up and to activate the cosmos. From these two postulates will ensue the thesis of La Mettrie: "Let us then boldly conclude that man is a machine, and that there is nothing in all the universe but a single substance diversely modified."[41] And by the same token, the decisive argument of the *Rêve de D'Alembert* summed up in the words: "If you concede that between the animal and you there is a difference only of organization, you will be showing sense and reason . . . but it will be inferred against you that with inert matter, heat and motion, one may obtain sensibility, life, memory, consciousness, passions, thought."[42]

In Bayle's period, certain *critiques* indited against the animal-machine not only mirrored clearly its progress towards naturalism, but might even have abetted unwittingly such a development. An example is Daniel's *Voyage du monde de Descartes*, previously examined, with which Bayle was familiar. The Jesuit's pointed analyses of Cartesian dilemmas seemed often to encourage a materialistic settlement of them. It was observed that Descartes's locating of the soul in the pineal gland, whence by the intermediary of the animal-spirits the organism as a whole was controlled, had in reality violated, in the physiological sphere, the dualistic basis of the metaphysics. In no sense could situation as such be made an intelligible mode or attribute of non-extended *res cogitans*. Whereas Malebranche and Leibniz had endeavored to solve

[41] La Mettrie, *Œuvres philosophiques*, III, 90.
[42] Diderot, II, 115.

this problem within the dualist framework, Daniel's discussion of the subject left only two logical inferences, both equally favorable to materialism. Either the soul was material; or matter possessed qualities proper to the mind.[43] As counterpart to his evaluation, Daniel remarked that, in regard to the *bête machine*, the Cartesians were at fault for doing things merely by halves;[44] that "every Cartesian, in order to speak consistently, ought also to say of other human beings who are in the world with him that they are automatons, just as seriously as he says it of beasts."[45] The Jesuit was, of course, trying to be satirical; but in Bayle's use of the same line of reasoning to insinuate materialist conclusions, there is proof of the fact that Daniel was actually wielding a boomerang.

The *Voyage du monde de Descartes* might, moreover, have helped to foster a naturalistic version of automatism by serving it up in a form that Diderot, for instance, could easily have utilized. Daniel considered, at some length, an objection made by orthodox Cartesians against the assimilation of man to beast implicit in the animal-machine. This was, namely, their insistence that language and an ideational sequence were, as Descartes had stated, evidence of man's essentially non-mechanical, rational nature. The Jesuit held that this did not quite demonstrate the point; for, a sequence in words and ideas might also be obtained *mechanically* from the premises of Cartesian physiology. Animals, although presumably only machines, did in fact outwardly exhibit both a succession of thoughts through the medium of their behavior, and the equivalent of language by their non-verbal means of communication.[46] Such an approach to the twin problems of

[43] Daniel, *Voyage du monde de Descartes*, p. 215.

[44] *ibid.*, 503.

[45] *ibid.*, 474.

[46] *ibid.*, 475ff. Daniel added that, according to Descartes, there ought to be less specific difference between man and the higher types of animal, than between the higher and lower animals: "Examinons bien icy toutes choses selon les Principes de la Philosophie Cartésienne, & nous verrons qu'il doit y avoir une bien plus grande différence entre la

knowledge and language seems related, in some measure, to the motivation of Diderot's interest in the non-lingual expression of deaf-mutes. The force of Daniel's parodic treatment of the *bête machine* was contained in his own incisive—and loaded—question: "What do you perceive in all these discourses and conversations other than movement?"[47] But the entire difficulty, once outlined, was dealt with in a persistently satirical vein:

"When we ask the reason for all these marvels, for this congruity and surprising diversity, we are told that they presuppose only the disposition of the machine, and then certain determinations of it which we are not able to perceive, but which come from exterior objects; that all these little machines determine themselves and variously modify one another. That is what the whole Cartesian doctrine boils down to. Why should I not be right in applying these two great principles, and these two solutions, to the behavior that those creatures called men manifest among themselves?"[48]

If Daniel was joking, the recurrence of the same argument in the *Pensées philosophiques* deserves to be appraised differently. While Diderot's dialectic in 1746 favored deism against the pretensions of the atheist, it is noticeable along what lines his subsequent conversion to naturalism was being prepared. In the following quotation, Cartesian automatism is at the core of a man-machine conception:

"Are you a thinking being? I asked him . . . Could you doubt it? he answered . . . Why not? What have I perceived

machine d'un ver de terre & celle des abeilles, afin qu'elles puissent faire entr'elles tout ce que nous y admirons, qu'il n'y en dévroit avoir entre les machines des abeilles & celles des hommes, pour faire produire à ceux-cy par les loix de la Méchanique cét arrangement de sons & de paroles que nous appellons discours suivi." *ibid.*, 494. The automatist viewpoint was thus capable of offering a support in physics to the chain-of-being concept adopted by eighteenth-century materialism.

[47] *ibid.*, 475. [48] *ibid.*, 492-493.

that convinces me of it? . . . sounds and movements? . . . But the philosopher finds as much in the animal that he deprives of the faculty of thought. Why should I grant to you what Descartes refuses to the ant? On the surface, you display actions that are likely to persuade me; I am tempted to assure that you do in fact think. But reason suspends my judgment. 'There is no essential connection between outward behavior and thought,' it says; 'possibly your antagonist cogitates no more than his watch. Must one take for a thinking creature the first animal that has been trained to speak? Who has revealed to you that men are not so many parrots instructed without your having knowledge of it? . . .' "[49]

The trend to materialism springing from Cartesian biology united, early in the eighteenth century, with libertinist and freethinking tendencies. It was in large part through this broader non-technical ambient that the radical outcome of Descartes's theories reached certain of the philosophes. Jean Meslier, whose writings were read in manuscript by the *incrédules* of Diderot's and La Mettrie's generation, gave naturalistic interpretations not only (as we have already seen) of the cosmogonic hypothesis in the *Traité du monde*, but of Descartes's *L'Homme* as well. The school of mechanistic physiology had claimed that an animal performed all its normal actions without benefit of an immaterial soul, or of any other intelligent principle fundamentally distinct from its corporeal organization. This suggested, by the standards of the *esprit simpliste*, an identical explanation for human conduct. Since Cartesians recognized, as Meslier pointed out, that "the circulation of the blood and the appropriate temperament of the humors cause the life of the body and all its motions, it is useless and absurd . . . to wish to imagine . . . without reason another principle of life, of which we have no need, inasmuch as the single principle that they ad-

[49] Diderot, *Pensées philosophiques*, pp. 14-15.

mit for the life of the body suffices both in us and in all other animals with respect to all the functions and actions of life."[50] The *Testament* turned its attention on the physiology present in Malebranche's *Recherche de la vérité*, and on his use of the *esprits-animaux* as a connecting link between organic operations and the workings of the will or intelligence. Meslier's rendering of Malebranchean ideas tended towards a denial of the role in man of an independent spiritual substance. Given the perfect correspondence and seeming interaction between the states of the soul and those of the body, there could be no valid excuse for doubting that thought and sentiment were in effect simply alterations of matter. "The Cartesians themselves confess that the diverse modifications and changes of the body excite in the soul diverse thoughts and sensations . . . and that there exists a natural relationship between these various modifications. . . ."[51] The unique problem remaining, therefore, was one that progressively deeper scientific inquiry alone could hope to solve—and solve to the profit of materialism: "in the opinion of those who maintain that the movement of matter and its different transformations are sufficient for giving knowledge and feeling to men and beasts, there is only . . . one difficulty present, namely, to know how this is done."[52] It was expected that anatomical and physiological research would by degrees clarify the causal factors behind the mind-body parallelism, and would dispel eventually the dilemma of "duplication" incurred by the dualist position. Denesle, surveying the scene of naturalistic philosophy around 1754, was to concede the soundness of Meslier's (and his successors') logic in the general resolution of the inconsistencies evident between Descartes's metaphysics and his theory of organism. About the growth of the *bête machine* into the *homme machine*, Denesle observed: "if the organs of the body necessarily admit of

[50] Meslier, *Testament*, III, 312.
[51] *ibid.*, 320.
[52] *ibid.*, 323.

sensation, and this latter of a soul, then animals who possess the same organs as we ought to have the same sensations, and consequently the same soul. . . . Now why should you accord to the human soul a destiny that you constantly refuse to the beast's? . . . Descartes, who renewed the singular and ill-founded opinion of the Spaniard *Gomez* [*sic*] concerning the material soul of animals, is no more consistent in his proofs than the other. Accordingly, these two systems have enjoyed an equal fortune, that is, no one except the Materialists have been in agreement with them."[53]

In Meslier's understanding of Cartesian biology, there is also present the conception that all organic phenomena are basically alike in that they derive uniformly from matter in fermentation. Such a view, supporting the notion of continuity in the multitude of living forms, worked towards the removal of the artificial barriers erected traditionally among the "reigns" of vegetable, animal, and human nature. Before culminating in the evolutionary materialism of Diderot and his associates, the idea of "organic uniformity" was notably reinforced by the Leibnizian chain-of-being hypothesis, and by certain pertinent discoveries in zoology during the 1740's. The conclusions stated by Meslier's *Testament* had shadowed forth the focal argument of a work such as La Mettrie's *L'Homme plante*, or the vitalist doctrine of the *Rêve de D'Alembert*: "It is necessary to affirm the same thing for all types of corporeal life, be it that of man, animals, or plants. Their life is nothing but a kind of modification or continual fermentation of their being, that is, of the matter of which they are composed."[54]

[53] Denesle, *Examen du matérialisme*, I, 383-385. The *Antoniana Margarita* (1554) of Gomez Pereira was regarded as the precursor of Cartesian automatism, owing to the Spanish physician's refusal of sentiment to animals. While it is unlikely that Descartes was directly influenced by Pereira, the latter's ideas were rescued from oblivion and set in vogue as a result of the subsequent formulation of the *bête machine*; cf., Bayle's article "Péréira" in the *Dictionnaire historique et critique*, XI, 546-564.

[54] Meslier, *op. cit.*, III, 339.

The proponents of Cartesian-inspired naturalism banished dualistic difficulties by identifying the *esprits-animaux*, or the "subtle matter" of which they were made up, with the soul itself. The *Traité des trois imposteurs*, among others, asserted that Descartes's pronouncements concerning spiritual substance revealed a confused position. The anonymous author came to the rescue, quite simply, by first admitting with Descartes that "there is in the universe a very subtle fluid, or a highly rarefied matter always in motion,"[55] and by then defining the soul, in terms of this material substance, as the rough equivalent of the animal-spirits common to man and beast.[56] Père Rapin had judged rightly that "the Physics of Descartes does not raise the mind above the level of the senses: it explains by means merely of bodies the most spiritual operations of the soul."[57] Subtle matter, represented physiologically by the *esprits-animaux*, served to obliterate the essential distinction assumed to exist by Cartesian metaphysics between mental and organic functions.

Contributing to the same end were the efforts of so renowned a medical authority as Boerhaave, with whom La Mettrie studied at Leyden, and to whom he owed much for the man-machine theory. La Mettrie himself informs us that, while Malebranche had given the *esprits-animaux* a preponderantly "philosophical" vogue, it was Boerhaave who did most to substantiate those claims experimentally.[58] The investigations of the Dutch physician, having their point of departure in the principles of the iatrophysical school, were directed, according to *L'Homme machine*, towards a materialistic definition of soul: "He explains by virtue of mechanism alone all the faculties of the rational soul. The most

[55] *Traité des trois imposteurs*, p. 86.
[56] *ibid.*, 87.
[57] Rapin, *Œuvres*, II, 431. The Jesuit critic was familiar with the inauspicious estimate of Cartesian natural philosophy given by More's *Enchiridion Metaphysicum*.
[58] La Mettrie, *Œuvres philosophiques*, I, 242.

metaphysical of thoughts, the most intellectual and eternally true, are all reduced by the famous theoretician to the laws of motion."[59] Denesle, writing when La Mettrie and Diderot were already being read, saw the animal-spirits as the chief material substance in which "the Materialists would like to have the essence of the human soul consist."[60] Before long, the abbé Coyer was to say of subtle matter, in a witty cataloguing of prevalent materialist ideas, that Descartes had employed it, like Empedocles his fire, to restore vitality, consciousness, and reason to the organs of the body. Coyer concluded: "Either there is no materialism, or that is it. And the public, who takes everything literally, reads all this."[61]

An important work reflecting the naturalistic outcome of Cartesian biology is Deslandes's *Pigmalion, ou la statue animée*. Although far from serious in tone, it anticipates with regard to both chronology and content the ideology of La Mettrie and Diderot. Deslandes's plan is to construct, in a manner recalling Descartes's "statue or machine" in the *Traité de l'homme*, a "living and animated statue: matter that passes through several stages, receiving different modifications, moving itself, acquiring sentiments. . . ."[62] The tale of Pygmalion, given an eighteenth-century turn, serves as the vehicle for Deslandes's principal philosophical concern, namely, the investing of a statue with the normal functions of life by means of a certain determination in its constitution: "I live, breathe, think, feel: could not the same occur in the case of this statue? All depends perhaps on a little more or a little less movement, and on a certain arrangement of its

59 *ibid.*, 265.

60 Denesle, *Examen du matérialisme*, I, 342-343.

61 Abbé Gabriel-François Coyer, "Lettre au R. P. Berthier sur le matérialisme," *Œuvres complettes de M. l'abbé Coyer*, Paris, 1782, I, 352.

62 André-François Boureau-Deslandes, *Pigmalion, ou la statue animée*, Londres, 1741, pp. ij-iij. Like La Mettrie's *Histoire naturelle de l'âme* and Diderot's *Pensées philosophiques*, the French authorities condemned the *Pigmalion* to be burned. Deslandes was well-known, in particular to Diderot, as the author of a standard *Histoire de la philosophie* (1737).

parts."[63] Vitality, sensibility, emotion are accordingly conceived as the effects of moving matter: "As movement is the medium through which matter, in order to change from non-thinking to thinking, must pass, the statue did not fail to acquire by degrees all the motions of which a body was susceptible."[64] Thus, Descartes's physiology found its final application; the mind was made essentially a dependency of the organism. The natural history of the soul coincided with that of the body: "at first it was nothing: it then becomes something and grows stronger; thereafter, it falls little by little into a state of non-existence, and is ultimately destroyed. Such is the life of the soul, which is hardly different from the body's."[65] The greater part of La Mettrie's argumentation in *L'Homme machine*, needless to insist, represents a more technical treatment of the same theme. Deslandes's metaphoric use of the statue to illustrate his point (indeed, Descartes's original device was to become quite familiar to the Enlightenment) suggests, moreover, a tie with the *Entretien entre D'Alembert et Diderot*, in which the first-named speaker asks: "I would very much like to know what difference you admit between a man and a statue, between marble and flesh." Diderot's answer to this is: "Not much difference. One makes marble with flesh, and flesh with marble."[66] At the same time, a hint of panvitalism is discernible in the *Pigmalion*: "Properly speaking, everything lives; and what appears to cease living, lives in another manner."[67]

63 *ibid.*, 30. 64 *ibid.*, 37-38. 65 *ibid.*, 37.

66 Diderot, II, 105.

67 Deslandes, *op. cit.*, pp. 45-46. The physiological naturalism of the *Pigmalion* was presented, not inappropriately, with admixtures of pantheistic metaphysics. One perceives in this the maturing effect of Spinoza's thought on the ideology that had sprung from Cartesian science: "Il y a des Etres sans nombre, qui existent tous à leur maniere, qui vivent & meurent tour à tour; mais tous ces êtres n'en composent qu'un seul, qui est le TOUT, qu'on appelle Dieu, la Nature & l'Univers. . . . Il y a apparence que le TOUT, que le vrai Etre doit contenir toutes les modifications possibles; & par conséquent il ne doit pas moins penser qu'être étendu, moins raisonner que se mouvoir, moins avoir des senti-

In 1743, one of many anonymous publications attesting the pervasiveness of freethinking, the *Nouvelles libertés de penser*, had occasion to discuss several of the foregoing topics. An article devoted to "The opinions of the Philosophers concerning the Nature of the Soul" quoted those persons who, to deny the principle of spirituality, had generalized Cartesian automatism from beast to man, from organic operations to intelligent behavior. Their reasoning followed a well-trodden path: "If you understand . . . what the cause is in animals of all their diverse functions, as well as how these take place, then you know, by supposing a higher degree of perfection in the organs on which such functions depend, the instrument and cause in man of rational thought."[68] Another essay in the collection, perhaps by Fontenelle, entitled "Treatise on Liberty by M* *" concluded that the soul's dispositions and actions are determined by those of the brain and nervous system; it thereby approximated both La Mettrie's automatist theory, and the moral determinism of Diderot and D'Holbach.[69] A third contributor dwelt, apropos of Descartes's employment of the pineal gland, on the manifest contradiction involved in placing spiritual substance within a material object and assuming that there could be interaction between them. Such a posing of the problem, it was declared, would plausibly lead only to a materialist interpretation of the soul's union with the body. For otherwise, the situation of *res cogitans* in non-thinking matter would by dualistic standards imply the impossibility of thought: "By supposing an intellectual substance united to a material

mens qu'être figuré." pp. 43-45. Analogous arguments are to be found in Diderot's critique of Maupertuis in the *Interprétation de la nature*, and in the *Rêve de D'Alembert*.

[68] *Nouvelles libertés de penser*, Amsterdam, 1743, p. 83.

[69] *ibid.*, 142ff. About the author of the "Traité de la liberté," the abbé Pluquet was to say: "Ses principes sur la nécessité des actions humaines, sont des conséquences de ceux du Cartésianisme sur l'union de l'ame & du corps." *Examen du fatalisme*, Paris, 1757, I, 415.

body, the destruction of intelligence would ensue from this union. It is necessary, therefore, to attribute to matter alone the operations commonly ascribed to spiritual substance, since the latter is incapable of them."[70] The *Nouvelles libertés de penser* is a good instance of the libertinist medium into which much of the naturalism deduced from Descartes's physiology was at first absorbed, before being made available to Diderot's group.

The *Recherches philosophiques* of the freethinker Saint-Hyacinthe bears witness, more plainly still, to the Cartesian sources of biological materialism in the years just prior to the debut of La Mettrie and Diderot. Saint-Hyacinthe reiterated the widespread opinion that Descartes's search for the seat of the soul in a bodily organ was, patently, a scientific activity at variance with his metaphysical position.[71] It was strongly insinuated that dualism, seeming prudently to compromise with theological dogma, had really been advanced with the expectation that certain *esprits-forts* would, in due time, resolve its dilemmas in favor of a naturalistic monism, as indeed La Mettrie was soon to pride himself on having done. The *Recherches philosophiques* said of Descartes:

"As soon as he established that the essence of matter consisted in extension and that all extended things were therefore material, Religion obliged him to say that the soul was not extended. . . . But those who affect a bit of Pyrrhonism . . . scoffed at the idea of a being which, having no extension, could not exist anywhere; or they suspected, on perceiving

[70] *Nouvelles libertés de penser*, pp. 164-165.

[71] Thémiseul de Saint-Hyacinthe, *Recherches philosophiques, sur la nécessité de s'assurer par soi-meme de la verité; sur la certitude de nos connoissances; et sur la nature des etres*, Londres, 1743, p. 493. Saint-Hyacinthe, a friend of the *libre penseur* Lévesque de Burigny, is a likely candidate for the authorship of the effective anti-religious tract, *Le Philosophe militaire*. Some *Pensées secrettes* appeared in 1735 under his name. Voltaire's attribution to Saint-Hyacinthe of his own *Souper du comte de Boulainvilliers* presumably did no injustice to the latter's reputation. Cf., Wade, *Clandestine Organization and Diffusion of Philosophic Ideas*, p. 51.

the contradiction present in the quest for its location, that Descartes had recognized its existence only in order to be safe with respect to theology. From then on, those so-called *Grands Esprits* thought themselves well-authorized to maintain that the soul was not an entity distinct from the body, but was simply comprised in the movements of the organic mechanism."[72]

Whereas Saint-Hyacinthe feigned disapproval of this materialist outgrowth of Cartesian science, the fact is that his airing of the issue was intended to propagate just such attitudes. Equally prejudicial to dualism was his obviously sly recommendation, reflecting a distorted Spinozistic viewpoint, to the effect that extension be made an attribute of spiritual substance in order to elude the dangerous pitfall prepared by Descartes. The immediate tangible result of the *Recherches philosophiques* is evidenced in its ironic remark that, given the perfect interdependence in man between body and soul, one would have to be thoroughly convinced by logical proofs to the contrary, or entirely penetrated by religious scruples, in order not to believe "that the soul and the body are the same thing."[73] Once this has been affirmed, it goes without saying that Saint-Hyacinthe neither adduces such proofs, nor excites such piety. His ideas must, moreover, have exerted a certain influence during the critical decade of the 1740's,

[72] Saint-Hyacinthe, *Recherches philosophiques*, p. 500. The "Grand Argument" of the Cartesian materialists is reproduced as follows: "L'Ame n'est point étendue, car elle seroit matiere. N'ayant point d'étendue, elle n'est nulle part. . . . Or on ne conçoit pas qu'un être qui n'est nulle part existe. L'Ame n'est donc rien qu'un resultat de la mechanique du corps, ce qui est si vrai . . . qu'en un mot l'Ame se sent de tous les divers états de bien & de mal où le corps se trouve. . . . Et non seulement la disposition où est le corps influe sur l'état de l'Ame, mais l'état de ce qu'on appelle *Ame* influe aussi sur l'état du corps. . . . Comment cela se peut-il faire . . . si l'Ame & le Corps sont deux Etres si differens que l'union en est inconcevable?" La Mettrie was familiar with the *Recherches philosophiques* and with the atheistic conclusion towards which, aided by Spinozist sympathies, Saint-Hyacinthe tended. La Mettrie, *Œuvres philosophiques*, I, 245.

[73] Saint-Hyacinthe, *op. cit.*, p. 508.

for they turned up in almost exact reproduction, before long, in the *Essay sur la nature de l'ame* of the physician Louis.[74]

Alluding to the materialism occasioned by Descartes's theory of organism, in 1738 Voltaire observed: "If beasts are pure machines, then you are surely, in relation to them, what a repeating-watch is compared to a turn-spit."[75] The great deist understood, at the moment, neither the precise import, nor the rich scientific possibilities of the notion he was ridiculing. When confronted at last with the positive outcome of automatist thought among his younger contemporaries, Voltaire felt compelled to defend a more traditional doctrine. By contrast, the marquis d'Argens had, as early as 1740, grasped the real significance for the age of the animal-automaton hypothesis, namely, its suggestion that matter itself might well be endowed with essential properties of sensibility and intelligence. While a theological precondition still clung to D'Argens's estimate, the apologetic application of the *bête machine* was wholly ignored: "I ask if it is not as likely that God should give perception to certain atoms and small corpuscles, as it is that machines should

[74] Antoine Louis, *Essay sur la nature de l'ame, où l'on tâche d'expliquer son union avec le Corps, & les loix de cette union*, Paris, 1747. After defining the organism as an assemblage of hydraulic and pneumatic machines, plus tubes and levers, Louis recognizes that an impasse has been reached as regards the discrepancies between, on the one hand, dualist metaphysics and, on the other, the physiological certainty of a "mutual dependence of the body and the soul" and the fact that there is "in the interior of our brain a fixed place where the soul resides." Avertissement, v-vj. His solution of the problem, purporting to be anti-materialist in aim, follows Saint-Hyacinthe in attributing extension to spiritual substance. But Louis, like his model, first advertises the materialists' obligation to Cartesian thought, and quotes in illustration of this their pernicious "Grand Argument." Whether sincere or not, his own metaphysical "reform," in rendering the soul essentially indistinct from *res extensa*, could only have corroborated the naturalism that he was supposedly combatting. One of the celebrated physicians of the period and a prolific writer on medical subjects, Louis was entrusted with the articles on Surgery for the *Encyclopédie*. Diderot knew him personally. IX, 410.

[75] Voltaire, *Traité de métaphysique*, crit. ed. H. Temple Patterson, Manchester University Press, 1937, p. 34.

be granted the power to act with the degree of wisdom they would have if actually possessing intelligence?"[76] Like Bayle, the *Philosophie du Bon Sens* interpreted the concept of a corporeal soul for beasts to the profit of human automatism: "If animals therefore have a material soul, sentiment is not incompatible with matter, but rather matter is susceptible to it. Who could deny that God, by purifying it and by rendering it more subtle, is able to raise matter to the level of understanding proper to the human soul?"[77] Consistently with the habit of physiologists who, from Descartes on, had envisaged man through the analogy of an abstractly assembled machine, La Mettrie claimed the seventeenth-century philosopher as an ideological forbear of his own thesis. He wrote: "A non-extended being cannot occupy space; and Descartes, who agrees with this truth, seriously looks for the seat of the soul and places it in the pineal gland. . . . What is the purpose, then, of the occasional causes by means of which the union of mind and body is explained? It is evident from the above that Descartes spoke about the soul only because he was forced to speak of it, and in the manner he did."[78]

The great vogue in France of Locke's passing conjecture that matter might be capable of thought must be ascribed, not to any exigencies of empiricist psychology as such, but rather to its having coincided with the materialism already extracted from the evolution of Cartesian science. Relying on Descartes's own logical criteria, D'Argens asked: "Is it not easier to believe that God grants thought to a substance that we know and of which we possess a clear and distinct notion, than it is to conceive of a substance having no extension, and about which we have no idea, that acts on matter?"[79] At the outset of *L'Homme machine*, La Mettrie

[76] D'Argens, *Philosophie du Bon Sens*, II, 226. This work first appeared in 1740.

[77] *ibid.*, 228.

[78] La Mettrie, *Œuvres philosophiques*, I, 240-241.

[79] D'Argens, *op. cit.*, II, 215.

cautioned against supposing on Lockean example that matter as such *thinks*, and advised instead that thought be regarded as the product of a certain mechanical activity on its part. The *Homme machine* was mirroring the historical consequences of Descartes's biology when its author prefaced his exposition with a dismissal of Locke's surmise: "It is plain from the first that we shall avoid this Scylla, where M. Locke had the misfortune of foundering."[80] When Diderot, in a well-known letter, urged upon Voltaire the materialist sense of the English philosopher's indecisive conjecture, this was not done primarily in connection with the sensationalist epistemology with which, for the most part, the *Lettre sur les aveugles* had been concerned. It was related, rather, to the interpolated episode of Saunderson's "vision of nature,"[81] which in turn represented an eighteenth-century offshoot of Descartes's *Monde*. Scholars have perhaps been overinclined to judge the philosophes' materialism as a general effect of empiricist thought. This would seem to involve, however, a serious error from the standpoint of intellectual history. While practically all the philosophes were students of Locke and rejected (overtly at least) Descartes's innatist psychology, it does not follow that La Mettrie, Diderot, D'Holbach, and Buffon were materialists *because* they accepted sensationalism. Although this last element was tightly integrated into their doctrine, there was no *decisive* tie, either logical or historical, between it and scientific naturalism. In fact, exponents of Scholasticism during the eighteenth century also advocated the empiricist theory of knowledge; and it was Condillac, perhaps the most implacably consistent critic of materialism at the time, who became the foremost Lockean among the philosophes. There was, as the current of ideas in the Enlightenment teaches, no necessary transition from Lockean philosophy to the natu-

80 La Mettrie, *Œuvres philosophiques*, III, 1-2.
81 Diderot, XIX, 420-421.

ralistic school here under scrutiny. But once the members of Diderot's group had derived the fundamentals of their materialism from the progress of Descartes's natural science, it was found that sensationalist epistemology conformed, on the whole, better with their position than did the belief in innate ideas. Innatism, as we shall see, was itself refashioned to harmonize with the mechanistic conception of man.

About the time that La Mettrie composed *L'Homme machine*, and possibly in some measure because of it, Diderot came under the sway of naturalistic Cartesianism. The *Bijoux indiscrets* of 1748, in seeking to define the soul in terms of organic functions, had a point of egress in automatist physiology. A personage who is apparently a Cartesian suggests, first, that the soul is situated in the head: "It is from there that the animal-spirits depart which, spreading by means of the nerves into all the other parts of the body, stop or move these latter according to the wish of the soul seated on the pineal gland."[82] The heroine of Diderot's novel, Mirzoza, expands this argument, bringing it presumably into line with more recent experimental data, when she contends that by the same logic one ought to say that the soul resides functionally, as occasion demands, in the various organs temporarily assuming predominance over the others.[83] The sequel to this half-facetious notion finds Mangogul, monarch of the fictitious kingdom of the *Bijoux indiscrets*, drawing inferences from Mirzoza's discourse that tend towards, if they do not actually formulate, the man-machine thesis. The flippant form in which Diderot's views are here conveyed should not detract from their importance. In Mangogul's opinion, it follows from Mirzoza's remarks that not only human beings have souls, but all beasts as well would be entitled to the same: "I have a soul; behold there an animal that acts most of the time as if it did not have one; and perhaps it does not have a soul, even when it acts as if it did.

[82] *ibid.*, IV, 247.

[83] *ibid.*, 247-250.

But it has a nose shaped like mine; I feel that I have a soul and think; therefore, this animal too has a soul and thinks. For a thousand years this argument has been tendered, and during all that while it has been merely an impertinence."[84] Even allowing for a kingly negligence about discursive style, Mangogul's meaning is made plain enough. Rather than claim (with the opponents of automatism) that animals must analogically be given a soul, he prefers to apply the analogy in reverse and to decide that man is no less a machine than the beast. "Far from agreeing with you," Mangogul replies to Mirzoza, "that everything possessing feet, arms, hands, eyes, and ears as I do possesses a soul like myself, I tell you I am convinced, and will never believe otherwise, that three-quarters of men and all women are only automatons."[85] These words are a clue to Diderot's intellectual development; his future writings will carry out and amplify the same philosophical theme in a less humorous vein.

In the *Entretien entre D'Alembert et Diderot*, it is said: "Do you pretend, with Descartes, that [the animal] . . . is a pure imitative machine? But children will laugh at you, and philosophers will answer that if it is a machine, then you are one too."[86] Diderot was himself one of those philosophers. If for him man was essentially a machine, however, the latter term, conforming with the career thus far described of Descartes's mechanistic physiology, was understood as an organism capable of life, sensation, and thought. D'Holbach recapitulated on that score the attitude of his colleagues: "man is an entity, resulting from the combinations of certain materials endowed with particular properties, whose arrangement is called the *organization*, and whose essence is to feel, think, act. . . ."[87] Like D'Holbach, Diderot followed La Mettrie in referring all human behavior to the physical organization and its mechanical operations. In that regard, the

84 *ibid.*, 251. 85 *ibid.*, 252. 86 *ibid.*, II, 115.

87 D'Holbach, *Système de la nature*, I, 11.

Entretien gave as good a summary of eighteenth-century naturalism as one may encounter in any text: "There remains in the universe but one substance, both in man and in the animal. The bird-organ is composed of wood; man, of flesh. The canary is made of flesh, the musician of differently organized flesh: but one and the other have the same origin, the same formation, the same functions, and the same end."[88] Diderot's vitalist materialism, in opposition to animistic theories of organism such as those of Stahl and his school,[89] expressed itself through the automatist conception that had called it forth. In the *Rêve de D'Alembert*, Diderot was prepared, speaking through Dr. Bordeu, to dismantle the foremost genius on earth "by a series of purely mechanical operations" into a mass of non-organized matter.[90] Descartes's principles in biology had eventually relegated the notion of soul to a secondary, and even incidental, role in the task of analyzing and defining man.

Up to Diderot's epoch, physiological and anatomical inquiry had served mainly to expand and confirm, by its progressive explorations, the general method devised by the *Homme de René Descartes* of setting up a theoretical parallelism between the human being and the machine. Concurrently with this trend, the special place of a spiritual substance in a self-winding, self-moving mechanism capable of life, sensation and intelligent behavior-patterns grew ever more recessive and ineffable. In the *Eléments de physiologie*, belonging probably to the 1770's, Diderot stated the final practical outcome of Descartes's philosophy of organism in relation to dualist metaphysics. Under the section-heading of "Animal et Machine," the beast-automaton idea reached in Diderot's unfinished treatise its logical culmination in a frankly naturalistic doctrine:

"What difference is there between a sensible and living

88 Diderot, II, 117. 89 *ibid.*, IX, 378. 90 *ibid.*, II, 177.

watch, and a watch made of gold, iron, silver, or copper? If a soul were attached to the latter, what effect would it produce therein? If the union of a soul to such a machine is impossible, let someone demonstrate it to me; if it is possible, let someone tell me what the results of such a union would be. The peasant who, seeing a watch move and unable to understand its mechanism, places a spirit in its hands is neither more nor less foolish than our spiritualists."[91]

Leaving aside as insoluble the problem of a metaphysical union between body and soul, Diderot thus felt warranted to restrain the latter's sphere of intelligible action to virtual non-existence. It was assumed, instead, that almost as much could be explained of human behavior by reference to purely mechanical causes, as Descartes had thought possible in the case of animals. Diderot accepted at face value the key argument of La Mettrie's *Homme machine*, to the effect that the soul's states and modifications were produced invariably by corresponding changes in the body.[92] Concerning the soul itself, the *Eléments de physiologie* said: "This spring, if it exists, is most subordinate. Its power is less than that of pain, pleasure, the passions, wine. . . . It is nothing without the body. I defy anyone to clarify anything without the body. Let someone attempt to explain how the passions enter the soul without corporeal movements. I defy him to explain it without starting from these movements."[93]

In the complex of the philosophes' materialism, the man-machine hypothesis occurred in close conjunction with a broad biological objective equally ascribable to the fund of Cartesian science. It was proposed to account mechanically and unfinalistically, not only for the total activity of the organic being, but for its origins and formation as well. The cosmogonical and geological setting for this phase of Cartesian influence on the Enlightenment has already been

[91] *ibid.*, IX, 265-266. [92] *ibid.*, 359-360. [93] *ibid.*, 377-378.

treated. It is necessary now merely to emphasize the obvious implications of the *Traité du Monde* for the progress of naturalistic biology. In the framework of Descartes's physics, the automatist theory of organism was, plainly, a logical adjunct of his plan to fabricate a world hypothetically, in its general *as well as particular* phenomena, from the laws of matter in motion. Rohault pointed out that anyone who accepted Descartes's views about the genesis of organic forms could not very well refuse to acknowledge the *bête machine* as being more than plausible: "For if one is attentive to the question, it is much more admirable that such a machine can be constituted than that it can walk or jump after it is made. Nevertheless, it is most certainly constructed, without either art or knowledge, by means solely of movement and the disposition of the particles of matter."[94] In like manner, the school of eighteenth-century materialism, specially as represented by Diderot, restored the man-machine to its fuller speculative context. The automatist thesis was integrated with the overall attempt to derive the cosmic order, in its inorganic superstructure as in its organic details, from the processes—whatever they might appear to be—of moving matter.

If in the *Monde* and *Principes* Descartes had failed to extend, as was his explicit intention, the account of the world's formation down to its minutest organic features, the reason was that at the time he had lacked sufficiently exact data concerning the animal organism. Descartes did not doubt, however, that the realm of biological phenomena comprised, without radical break, a single unit with the non-living universe. It ensued therefrom that the conventional dichotomy between the organic and the inorganic was simply a nominal or formal, but not a real or material, one. Writing to Mersenne (who despite his apologetic interests had a somewhat naive and uncritical faith in mechanistic philosophy), Descartes had voiced firmly the scientific hope of giving a funda-

[94] Rohault, *Entretiens sur la philosophie,* p. 164.

mentally naturalistic explanation of the emergence of organic beings. A theological reservation, which a century later was no longer wholly to satisfy the foes of materialism, accompanies the philosopher's statement of purpose:

"The multitude and order of the nerves, veins, bones, and other parts of an animal, do not at all indicate that Nature is not adequate to form them, provided it is supposed that Nature acts in every case according to the precise laws of mechanics, and that God has imposed these laws on her. . . . I have found no thing whose particular formation I do not think myself able to explain by natural causes, just as I have explained in my *Meteors* that of a grain of salt, or of a small flake of snow. And if I were to start my *Monde* all over again, in which I have assumed the body of an animal to be already formed and have satisfied myself with showing its functions, I would undertake to give also the causes of its formation and of its birth."[95]

After a decade of labor on the project, Descartes informed another (unidentified) correspondent around 1649 that, although he had previously almost renounced the ambition of explaining the origins of animals, "in thinking thereupon, I have discovered so much new ground that I hardly doubt I will be able to complete the whole of Physics according to my wish, provided I have the leisure and facilities needed for making a few experiments."[96] Descartes's premature death left unfulfilled the biological prospects of his natural philosophy. Not until the age of *lumières*—and then principally by Diderot himself—was the Cartesian enterprise recommenced and put to execution in the comprehensive spirit of its author.

Descartes's exclusion of teleology from the province of the sciences of life had struck at the very roots of an entire composite of traditional beliefs. This fact, perhaps not too

[95] Letter dated 20 Feb. 1639, in Descartes, II, 525.
[96] *ibid.*, v, 261.

apparent at the start, was to be made steadily clearer by the workings of ideological history. That Cartesian thought was, on the point in question, immeasurably nearer the attitudes of Diderot's group than was, for example, the diluted Epicureanism of Gassendi, may be learned from the latter's defense, against his contemporary, of final causes, particularly as these applied to organic patterns. Answering the criticism of Gassendi's *Cinquièmes Objections*, Descartes made his own position fairly transparent. While there was a certain ineffable sense in which it would be true to say that God, by the inscrutable decrees of Providence, had the ultimate responsibility for the Creation; and notwithstanding the pious utility of conjecturing about the divine purposes manifested in nature, it remained constant that such theological perspectives "certainly in physics where all things must be supported by solid reasons . . . would be inept."[97] One could hardly have been more outspoken in the seventeenth century. In 1678, Cudworth classed Cartesian science on that score with the leading modern successors to Epicurean materialism, holding that the transference of purely mechanistic principles from inanimate to animate matter would lead invariably to atheism.[98] During the same period in France, and corresponding with the "christianizing" efforts of certain of his disciples, a general relapse from Descartes's biological goals took place. Malebranche, who had found it arduous enough to reconcile mechanistic physiology with spiritualist dogma, refused to subscribe further to the plan of elucidating unfinalistically the generation of organic beings. He earned thereby the felicitations of Leibniz, who entertained serious misgivings apropos of Descartes's dismissal of finality.[99] But such providentialist

[97] *ibid.*, VII, 309.

[98] Cudworth, *Intellectual System of the Universe*, II, 94-95, and *passim*. Cf., also, Emile Boutroux, "Descartes and Cartesianism," in *Cambridge Modern History*, IV, 791.

[99] Leibniz wrote to Malebranche (13 Jan. 1679): "Je trouve aussi que

leanings on the part of certain heirs to Cartesian philosophy came to signify, in the eyes of the eighteenth century, a distortion and violation of Descartes's intent and procedure in biology. Montesquieu, talking in particular about botany, made this altogether plain in his "Observations sur l'histoire naturelle" of 1721.[100] Similarly, an adversary of the philosophes such as the abbé Pluche, in examining Descartes's contribution to the growth of materialism, remarked that the "official" Cartesians, despite their willingness to indorse a good portion of the *Monde's* cosmogenetic speculations, had broken with the master's doctrine of organic origins by their return to teleological concepts.[101] By contrast, it was the "unofficial" current of Cartesianism that brought to fruition this special phase of the philosopher's system of nature. Pluche himself stressed the contradiction in ideas of those Cartesians who, while admitting that vaster cosmic phenomena resulted from the general laws of matter in motion,

vous faites un très-bel usage des causes finales, et j'ai une mauvaise opinion de M. Descartes qui les rejette, aussi bien que de quelques autres de ses endroits où il montre son âme à découvert." See Victor Cousin, *Fragments de Philosophie cartésienne*, Paris, 1852, p. 372. While advocating *tourbillon* physics, Malebranche had arrested the materialist tendency of Cartesian science at the question of biological development; he explained the formation of organisms by virtue of an "emboîtement des germes" theory which reintroduced finalism into the natural order.

100 "Ceux qui soutiennent que les plantes ne sauroient être produites par un concours fortuit, dépendant du mouvement général de la matière, parce qu'on en verrait naître de nouvelles, disent là une chose bien puérile. . . . Nous finissons cet article par cette réflexion, que ceux qui suivent l'opinion que nous embrassons peuvent se vanter d'être cartésiens rigides, au lieu que ceux qui admettent une providence particulière de Dieu dans la production des plantes, différente du mouvement général de la matière, sont des cartésiens mitigés qui ont abandonné la règle de leur maître." Montesquieu, *Œuvres*, VII, 47-48.

101 Pluche, *Histoire du Ciel*, II, 248: "Ici la division se mèt entre le maître & les disciples. Descartes qui nous a promis de faire sortir de ses trois élémens les choses *particulières comme les générales*, veut, bon-gré mal-gré, nous fournir encore la mer & les poissons. Mais ses disciples l'abandonnent & me répondent unanimement, que quand il s'agit d'espéces organisées il faut changer de principes, & recourir à des plans particuliers, & à des volontés spéciales."

wished to make organic facts dependent on final causes.[102] But the Enlightenment could hardly have been in the least ignorant of the *Monde's* actual objectives. In that regard, the *Encyclopédie* summarized correctly the sense of Descartes's physico-biological hypothesis: "The same mass of dust which has furnished us with an earth, planets, and comets arranges itself by virtue of motion into other forms, and gives us water, an atmosphere, air, metals, stones, animals, and plants; in a word all things, *both general and particular, that we see in our world,* organic as well as inorganic."[103]

Descartes's biology was held partly and temporarily in check, during the first half of the eighteenth century, by a profusion of writings on natural theology (Pluche's *Spectacle de la nature* was perhaps the most influential in France), all of which sought to interpret organic forms as proofs of final cause and providential design in nature. Nonetheless, Maupertuis's *Essai de Cosmologie* recognized, towards 1750, that there existed already two main, and seemingly irreconcilable, orientations in science: the finalistic, and the naturalistic. Comparing the two, he said of those who advocated the latter: "Too little affected by the marks of intelligence and purpose that one finds in Nature, they would like to banish from it all final causes. One group sees the supreme Intelligence everywhere; the other sees it nowhere. It believes that a blind mechanism has been able to compose the most highly organized bodies of plants and animals, and has succeeded in working all the marvels which we observe in the universe." At this point, a footnote by the author refers the reader specifically to the *Principes* and *L'Homme* of Descartes.[104]

102 *ibid.*, 249, 250ff.

103 Diderot, D'Alembert, *et al.*, *Encyclopédie*, II, 723; article "Cartésianisme," by the abbé Pestré.

104 Maupertuis, *Œuvres*, I, 13. Elsewhere, Maupertuis made known the same conflict when he juxtaposed the "emboîtement des germes" theory

But the teleological faction itself, as Maupertuis well realized, had by its exaggerations fallen into a kind of manifest absurdity during the 1740's. Not only was its method barren or futile, and sometimes even pernicious, in the search for the determinants behind organic facts, but it perpetrated logical confusion in face of the evidence of imperfections or contradictions abundantly present in the natural order, for which of course no final cause or supernatural design could reasonably be assigned. Therefore, certain thinkers found it all the more necessary to conceive of Nature, i.e. the *ensemble* of physical particles and the laws governing their complex motions, as possessing self-directing, creative powers. Descartes himself, it appears, had foreseen to what conclusions the problem of finalism *versus* naturalism might eventually lead: "Someone will say disdainfully that it is ridiculous to attribute so important a phenomenon as the formation of man to such inferior causes. But what greater causes are required than the eternal laws of nature? Does one demand the immediate intervention of an intelligence? Of which intelligence? Of God's? Then why are monsters born?"[105] About a century later, the *Lettre sur les aveugles* put the same argument, more dramatically, in the mouth of Saunderson, who was himself one of the "monsters." Unable to find any divine intercession in his having been born blind, Saunderson preferred to relate, for motives both of logical coherence and of true piety, the origination of organic beings, and of the cosmos as a whole, to "matter in fermentation." Reflecting by his own physical condition the corresponding degree of cecity in the operations of Nature, Saunderson was in 1749 the appropriate herald and spokesman for Diderot's scientific naturalism.

with the "efforts de Descartes, & de quelques-uns de ses disciples, pour expliquer par la seule étendue & le seul mouvement la formation des animaux & de l'homme." *ibid.*, II, 144.

105 In *Primae cogitationes circa generationem animalium*, Descartes, XI, 524.

The Cartesian background of the philosophes' materialistic biology has thus far been traced in its broad outlines. But to understand why, and in what form, the latent possibilities of Descartes's science triumphed at long last in the age and school of Diderot, it is necessary to consider their historical impact in combination with the acquisitions and exigencies of the vitalist trend that took hold of biological interest in the 1740's. This latter development was variously exemplified by Trembley's discovery of the "polyp," Bonnet's disclosure of parthenogenesis, Needham's attempted rehabilitation of the doctrine of spontaneous generation, Haller's demonstration of irritability and sensibility as properties inherent to muscle fibre, Buffon's theory of organic molecules, and other connected investigations or problems. It served gradually to render obsolete the view of a created or passive order of things, and to put in its place the crucial notion of a creatively active Nature. This ideological transition occurred irresistibly; its end-result was not greatly affected by the efforts of the many naturalists who, while contributing to vitalistic materialism, persisted in evaluating their findings in accordance with the traditional teleology. The naturalism of Diderot and his colleagues was, in large measure, the philosophical expression of the new conception of Nature insofar as it rested on vitalist proofs. But such a conception, implicit from the beginning in Cartesian physics, had already enjoyed a continuous growth up to the period of Diderot. It was to be expected, then, that the separate streams of Cartesian materialism and of vitalist biology, merging in the course of the 1740's, would mutually strengthen each other and bring to maturity a doctrine that broke definitively with the past. If Trembley's polyp, Needham's experiments with Infusoria, and like subjects, promptly became the basis of materialist speculation that ranged far beyond the limits of zoology, the reason was that certain philosophes were prone to appreciate the significance of such scientific events in the light of the radical lessons deduced from Cartesian thought. In turn, the consequences of the vitalistic current in Diderot's *milieu* fell, philosophically, into the ready mold of Descartes's mechanistic and unfinalistic biology. If this should, on the surface, appear paradoxical, one might do well to see how it happened.

Trembley's fresh-water hydra had the amazing ability, when cut up into pieces, of regenerating into as many complete new polyps as there were severed portions of the original one. The polyp exercised a deep and in some ways decisive influence on the genesis of eighteenth-century materialism, in particular with regard to La Mettrie.[106] In the context of vitalist science at the time, Trembley's famous zoophyte came to be the most widely-bruited and convincing clue to the immanence of self-determining powers in matter as such, or of the capacity of Nature to form organic beings by its own inherent laws without the imposition of design from the outside. It was by reference specifically to the polyp's properties that La Mettrie, in *L'Homme machine*, formulated the idea of a Nature, which was neither God nor Fate, to whose hidden but necessary causes the shaping and running of the universe, as of all its varied phenomena, might be ascribed.[107] In doing so, La Mettrie's materialism transcended the boundaries of Epicurean philosophy with its meaningless concept of Chance, and carried out instead

106 Cf., my "Trembley's Polyp, La Mettrie, and 18th-century French Materialism," *Journal of the History of Ideas*, Vol. xi, no. 3 (June 1950), pp. 259-286.

107 La Mettrie linked this naturalistic assumption with the proper scope and aim of science: "quelle absurdité y auroit-il donc à penser qu'il est des causes physiques pour lesquelles tout a été fait, & auxquelles toute la chaîne de ce vaste univers est si nécessairement liée & assujettie, que rien de ce qui arrive, ne pouvoit pas ne point arriver, des causes dont l'ignorance absolument invincible nous a fait recourir à un Dieu. . . . Ainsi détruire le hasard, ce n'est pas prouver l'existence d'un être suprême, puisqu'il peut y avoir autre chose qui ne seroit ni hasard, ni Dieu, je veux dire la nature, dont l'étendue par conséquent ne peut faire que des incrédules; comme le prouve la façon de penser de tous ses plus heureux scrutateurs." *Œuvres philosophiques*, III, 54-55.

the implications of Descartes's definition of Nature in terms of a fixed, determinable mechanism.

Not only was the polyp behind the view of a creative Nature that, by virtue of its innate laws, decided its own development and order, but Trembley's discovery offered, also, striking confirmation of the *bête machine* theory considered in its materialistic sense. In the history of the animal-automaton controversy, Descartes's notion of corporeal soul as something substantially indistinct from matter had been opposed and rejected by a number of critics and polemicists. These latter, by way of refuting the automatist thesis, had maintained the existence in beasts of what they called, by a curious abuse of terminology, "material soul." As an adaptation of the Scholastic *anima sensitiva*, the material soul was expected in somewhat hybrid fashion to account for sensation and thought in animals without itself being either properly spiritual or quite the same as matter, but instead something halfway between the two extremes. This argument, incomprehensible as it seemed to many, proved to be a valuable asset of the anti-automatist camp. Against it the Cartesians objected that material soul must be understood as identical with what Descartes had designated as corporeal soul, that is, the bodily mechanism itself. From the Cartesians' standpoint it ensued that corporeal soul, unlike the indivisible material soul, would have to be divisible like all matter if, in reality, it consisted uniquely of that substance. But here an *impasse* was reached, for it could not be shown that the vital principle and sensibility manifested in the organism, and included in the definition of corporeal soul, were in actual fact divisible, and that the soul was in essence extension. Mirrored through Diderot's mind, this was to become the question of whether sensibility (taken as a mechanical function) was or was not a general property of matter. Meanwhile, recapitulating the state of the conflict between Cartesians and non-Cartesians, D'Argens said:

"Thus, the objection made that the soul, being corporeal, must be divisible in several parts each of which would be a soul, is not borne out, not even in animals; for their divided and severed parts are but simple small corpuscles that no longer have any sensation or knowledge."[108] The problem of the nature of animal-soul being set up in this form, it is obvious that the discovery of the polyp, whose "soul" was observed to be as indefinitely divisible as Cartesian matter, could not fail to prove conclusively for the materialist-automatist doctrine.[109] This circumstance explains, in large measure, the philosophical vogue enjoyed by Trembley's hydra in certain circles.

The comprehensive result of discussions based on such phenomena as the polyp's regeneration, muscular irritability, and the like, found embodiment in the pages of *L'Homme machine*, whose thoroughgoing materialism may be said thus to have coincided with the theoretical impact of vitalist biology. The triumph of the new approach to organic facts and of the old Cartesian automatism was achieved by mutual collaboration. This may best be noted by comparing the *Homme machine* with La Mettrie's earlier *Histoire naturelle de l'âme* (1745).[110] In the last-named work, where the influence of the polyp's powers and of muscular irritability was still lacking, La Mettrie's "materialism" (if it can be

108 D'Argens, *Philosophie du Bon Sens*, II, 243-244.

109 The mathematician Cramer wrote to Bonnet, who took a leading part in speculation about the polyp: "Il est vrai que vos observations portent un rude coup au système de l'âme des bêtes; mais je ne dirai pourtant pas encore que tout soit désespéré. Une docte ignorance pourrait bien être . . . la conclusion de tous nos raisonnements. Au pis-aller, si ce ne sont que des machines, elles n'en sont que plus admirables. Car vous ne leur donnez une âme que parce que vous expliquez plus aisément par là leurs actions qui paraissent libres et tendre à certaines fins. S'il était bien prouvé qu'elles n'ont point d'âme, il faudrait renoncer à cette explication si simple, si naturelle; les choses ne se passeraient pas avec tant de facilité que nous le voyons." Quoted in Caraman, *Charles Bonnet*, Paris, 1859, pp. 30-31.

110 Cf., my "Trembley's Polyp, La Mettrie, and 18th-century French Materialism," pp. 275ff.

called that at all) had remained somewhat incoherent. Its effectiveness was mitigated, moreover, by his primary concern, not with mechanistic philosophy, but with an empiricist theory of knowledge, and by his reliance on concepts of vegetative and sensitive soul, as well as on the Scholastic "substantial forms," to explain organic process—all of which served, in 1745, to inhibit the creativity of Nature. Simultaneously with this, the *Histoire naturelle de l'âme* had ridiculed the beast-automaton idea. But in the *Homme machine*, by a singular *volte-face* La Mettrie's dominant interest, and with it his philosophical frame of reference, shifted from Lockean psychology to mechanistic physiology. He discarded entirely all vestiges of Aristotle's biological doctrines, and admitted that on the issue of animal automatism his opinion had, since previously writing, undergone a radical change in favor of the Cartesian position.[111] It is in the light of these events that La Mettrie's claims of historical filiation, by which in 1748 Descartes was made responsible for the man-machine, deserve to be evaluated.

Between 1746 and 1749, Diderot's gradual turning away from the deism of the *Pensées philosophiques* to the plain-spoken naturalism of the *Lettre sur les aveugles* was controlled by the advance of vitalist biology, and by the bold conclusions to which, as in La Mettrie's case, it was apt to give rise.[112] The inferences drawn from the polyp's self-generative habits, absorbed before long with Needham's proofs of spontaneous generation into Buffon's theory of *molécules organiques*, were to pass finally into the philosophy of organism expounded by the *Rêve de D'Alembert*.[113] Meanwhile,

[111] La Mettrie, *Œuvres philosophiques*, III, 79.

[112] See my "From Deist to Atheist: Diderot's Philosophical Orientation, 1746-1749," *Diderot Studies*, pp. 46-63.

[113] See Diderot, II, 124-130, where the polyp's characteristics are generalized to the extent of suggesting how the contiguous assembling of "sensitized particles" could have formed, without reference to finalism, organic phenomena, animal species, etc. In a letter to Sophie Volland (15 Oct. 1759), Diderot expressed the nexus in his thinking between the

Trembley's findings prompted a whole series of investigations, by some of the foremost naturalists of the period, into analogous phenomena of regeneration and resuscitation. In Diderot's thinking, the mass of such vitalist evidence became intimately connected with both the man-machine hypothesis and the naturalistic view of organic origins. This is clearly (and wittily) perceived, for instance, in the *Entretien d'un philosophe avec la maréchale de* * * *, where Crudeli is allowed to speak in the author's stead:

"La Maréchale: But this world, who has made it?

Crudeli: I ask *you*.

La M: God has.

Cr: And what is God?

La M: A spiritual being.

Cr: If a spirit can make matter, why is it that matter cannot make a spirit?

La M: Why should it?

Cr: The fact is that I see it doing so every day. Do you believe that beasts have souls?

La M: Of course I do.

Cr: Then could you tell me what happens, for example, to the soul of the Peruvian serpent which, while hanging in the fireplace and continually exposed to smoke for a year or two, dries up?

La M: Whatever should happen to it, what difference is it to me?

Cr: The point is that Madame is not aware that this smoked, dried-up snake resuscitates and is reborn.

La M: I don't believe a word of it.

isolated case of the polyp and the inherent creativity of nature as a whole: "Nous jugeons de la vie des éléments par la vie des masses grossières. Peut-être sont-ce des choses bien diverses. On croit qu'il n'y a qu'un polype! Et pourquoi la nature entière ne serait-elle pas du même ordre? Lorsque le polype est divisé en cent mille parties, l'animal primitif et générateur n'est plus; mais tous ses principes sont vivants." *ibid.*, xviii, 409.

CR: Nonetheless, it's a very able man, Bouguer, who assures this to be so.

LA M: Your able man has lied.

CR: And what if he has told the truth?

LA M: Then I would be obliged to think that animals are machines.

CR: And man too, who is simply a slightly more perfect animal than the others. . . ."[114]

The manifestations of vitalism in nature, by identifying experimentally the principle of life and feeling with the functions of a machine-organism, facilitated the application of the beast-automaton notion to man himself. But more than this, the new biological trend confirmed the belief of certain philosophes that Nature was itself answerable for the emergence of its own order, and for its patterns both organic and inorganic. A specially important instance of such an influence is the work of John Turberville Needham. Having pursued several lines of inquiry suggested by the polyp's recreative properties, Needham had rehabilitated in the eyes of many the doctrine of spontaneous generation. It was claimed, on the basis of numerous experiments with infusions of vegetable and animal substances, that putrefaction was capable of producing worm-like micro-organisms. Needham's *Observations microscopiques* (translated from the English in 1747), which gave detailed accounts of his investigations, was destined to be a leading factor in the shaping of Diderot's naturalism.

Needham's ideas had appeared to reverse completely the doctrine of generation held almost without exception by the scientists of his time. Since Redi had proved in the seventeenth century that flies did not result from deteriorating matter, the case for spontaneous generations had been without any empirical support. Having proscribed with this Des-

[114] *ibid.*, II, 520-521.

cartes's project to explain mechanically the origins of living things, traditionalist opinion—from Hartsoeker, through Malebranche, to Voltaire—had up to Diderot's period felt secure in the belief that all organic productions occurred by means of *germes*. These "seeds," viewed theologically, were put forth as unmistakable repositories of special providential designs, and as evidence of finality in nature. The deism of Diderot's *Pensées philosophiques*, with its roots in Newtonian teleology, had found corroboration in the scientific statement that "all observations concur in proving that putrefaction alone does not engender any organized beings."[115] But Needham's "anguilles," as they came to be called, transposed all this in Diderot's thinking. Also, they accredited inevitably the broad aims of Cartesian biology by implying that the causes of all organic—and *a forteriori* of all inorganic—developments were to be looked for, not outside, but within Nature. The realization of Descartes's unfinished system of physics became an exigency of the philosophes' science. A scheme of explanation was needed that would unfold the process by which matter had from its own properties and laws of motion constituted the world in all its intricate phenomena, "tant générales que particulières." In the *Lettre sur les aveugles*, the notion of spontaneous generation, fused with a transformistic hypothesis, was promptly magnified by Diderot to cosmic proportions, and the focal point of his entire vision of nature became the question of how, "in the beginning . . . matter in fermentation caused the world to take shape."[116]

The fact that Needham's data could quickly become, in the hands of certain philosophes, the pretext for embarking

[115] Diderot, *Pensées philosophiques*, p. 13: "La seule découverte des germes a dissipé une des plus puissantes objections de l'Athéïsme." The critics of naturalism at the time concurred, whatever their differences otherwise, in the axiomatic belief that organic forms were the products of *germes* embodying special purposes imposed on matter externally.

[116] Diderot, *Lettre sur les aveugles*, p. 43.

on a speculative account of cosmic origins, including even man, is itself highly symptomatic of how central to the materialism of Diderot's group were the broad prospectus and audacious objectives of Descartes's *Monde*. Attesting to a deep-seated sympathy during the Enlightenment for *a priori*, rationalist method, Needham's experimental findings were freely and imaginatively generalized in the *Rêve de D'Alembert*. There they furnished a cosmogonic conception which, together with overtones of transformism, was the continuation and analogue in the eighteenth century of the Cartesian universe of worlds in successive destruction and reconstruction. A portion of D'Alembert's dream is described as follows:

"Then he started to mutter I know not what about grains, about pieces of flesh macerated and placed in water, and about different species of animals that came one after the other into being and passed away. He had imitated with his right hand the tube of a microscope, and with his left, I believe, the open end of a flask. He looked into the flask through the tube, and said: 'Voltaire can joke about it as much as he likes, but the man-with-the-eels is right. I believe my eyes. I see them. How many there are! . . .' The flask in which he perceived so many momentaneous generations taking place he compared to the universe. He saw in a drop of water the history of the world. This idea appeared far-reaching to him; he found it altogether in agreement with sound philosophy, which studies large bodies through the small. He said: 'In Needham's drop of water, everything is executed in a moment. In the world, the same phenomenon takes a little longer. But what is our duration contrasted with the eternity of time? Less than the drop I have taken with the needle's point in comparison with the limitless space that surrounds me! There is an indefinite series of animalcules in the fermenting atom, and the same indefinite series of animalcules in that other atom called the earth. Who knows what species of animals have preceded us? Who knows what species will follow those of our epoch? All things change, all passes, only the whole remains. The world commences and finishes continually. . . .' "[117]

Voltaire, who persevered in the teleological view of a passively created nature, had eyed with distrust all signs of spontaneity or self-regulating powers in matter. Accordingly, he had tried to efface by satire the result of Needham's labors. Needham himself, in fact, had repeatedly disowned publicly the materialist theorizing for which his discoveries had provided both an occasion and an impetus.[118] But if, despite such restraints, the *Observations microscopiques* played a most serious role in the maturing of scientific naturalism, this event (otherwise puzzling, perhaps, and out of joint with the book's limited scope) may be clarified in terms of the ideological environment of the 1740's. Needham's data, like other happenings in science at the time, had actually corroborated and reinforced a strong, but subterranean current of thought that required little encouragement to erupt to the surface. On the urgings of that current, Diderot's belief in the creativity of Nature became fixed by a kind of logical necessity. Even after Spallanzani disproved in 1768 the claims advanced from Needham's experimentation with putrefying wheat-grains, the affirmation of spontaneous generation was to be left unmodified both in the unpublished text of the *Rêve de D'Alembert*, and in the random notes and reflections of varying date collected as the *Elèments de physiologie*,[119] to which, it is known, numerous additions were made during the 1770's. Voltaire was perfectly aware of how neatly Needham's investigations had fitted in with the general postulates and goals of scientific

[117] Diderot, II, 131-132.

[118] I have in preparation a separate study of the general role of Needham's *anguilles* in the intellectual history of the period.

[119] Diderot, IX, 263.

naturalism: "A physicist [who could be La Mettrie, Diderot, Buffon, or D'Holbach] . . . did not doubt that this Needham was a profound atheist. He concluded that since worms could be produced with rye-flour, human beings could be made from wheat-flour; that nature and chemistry brought about all things, and that it was demonstrated that one could do without a Deity who forms all creatures."[120] D'Holbach, in particular, cited Needham's discovery in support of materialism by combining it with the automatist consequences of Descartes's iatrophysical biology: "In moistening flour with water and sealing up this mixture, it is found after a while, with the aid of the microscope, that organized beings have thereby been produced that enjoy a life of which flour and water were considered incapable. It is thus that inanimate matter is able to pass to life, which is itself only an assemblage of movements."[121] On the basis of this fact—which Voltaire, incidentally, took to be the cornerstone of D'Holbach's philosophy, the error "on which he has built his system"[122]—the *Système de la nature* asked, in a footnote, if "the production of a man, independently of the ordinary means, would be more marvelous than that of an insect with flour and water?" This touches upon the problem of transformism, which will be examined presently.

Buffon's interpretation and use of the data in question was a bold step towards carrying out Descartes's project of explaining mechanically and unfinalistically the formation of organisms. On the evidence of spontaneous generation in Trembley's polyp and Needham's "eels," Buffon was led, in the introductory portions of the *Histoire naturelle* appearing in 1749, to conceive that there was dispersed in nature an indeterminate quantity of "organic molecules." These, by joining together according to regular and intelligible laws

[120] Voltaire, *Dictionnaire philosophique*, I, 397.
[121] D'Holbach, *Système de la nature*, I, 24-25.
[122] Voltaire, *op. cit.*, I, 397.

of mechanics, were held to be responsible for the production of all living phenomena in a manner analogous to that of mineral substances.[123] Buffon's vast and imaginative theory, wholly discarding final causes in the biological sphere, appeared to be fundamentally a realization of what Descartes had not lived to execute according to his wish. Giving to vitalist materials an evaluation that suited his aims, Buffon believed that he had "proved by factual means that matter tends to organize, and that there exists an infinite number of organic particles; I have therefore simply generalized from observation, without having advanced anything contrary to mechanical principles."[124]

Before discussing the bearing of Buffon's work on transformistic materialism, it would not be amiss to note how the *molécules organiques* renewed, in the Enlightenment, the specifically Cartesian doctrine of reproduction and its ideological connotations. It was well-known to the philosophes that Descartes had proposed a somewhat crude theory of chemico-mechanical epigenesis: "... man was formed from the mixture of the seminal liquids of both sexes. This great philosopher, in his *Traité de l'homme*, believed that he could explain, by the laws of motion and fermentation alone, the formation of a heart, brain, nose, eyes, etc."[125] The second half of the seventeenth century, however, saw the almost complete sacrifice of Descartes's line of inquiry to the preformationist viewpoint. The belief that Nature, far from creating new life, merely developed organisms already given in their seeds was, in turn, exploited theologically, and this came to represent simply another phase of the apologetic tradition into which official Cartesianism relapsed. Swammerdam expanded before long the preformationist position into a theory of *emboîtement des germes*, which was incorpo-

123 Buffon, *Œuvres*, I, 434-438.
124 *ibid.*, 453.
125 Maupertuis, *Œuvres*, II, 67. Cf., article "Génération" of the *Encyclopédie.*

rated by Malebranche into the *Recherche de la vérité.* This conception, claiming that all living things, past, present, and future, pre-existed, encased indefinitely one within the other as far back as the first members of each species, made it possible to attribute all organic effects finalistically to a single initial Creation. Leeuwenhoek's discovery of spermatozoa soon divided scientific opinion into the two main schools of ovists and animalculists, but this of course did not lessen the teleological efficacy of preformationism. Against such a background, the problem of generation became a touchstone to the ideological conflicts of the eighteenth century. Among the philosophes, the preexistence of *germes* was defended, despite differences in detail, by Voltaire, Réaumur, Bonnet, and Haller as an essential tenet of deistic or Christian faith, and as a bulwark against naturalistic biology.

By the same token, the case for epigenesis, dormant since Descartes, became an important concern of materialist science. Influenced by vitalist facts, and eager to account for bilateral heredity and for hybridization (for which preformationists promised no satisfactory solution), Maupertuis reintroduced the epigenetical hypothesis to the age of *lumières* with his strikingly original *Vénus physique* of 1745. He recognized that "Descartes's sentiment concerning the formation of the foetus . . . has something remarkable about it."[126] However novel Maupertuis considered his own explanations to be, he was aware of starting out from more or less the point where Cartesian physics had left off.[127] To remedy the deficiency of the mechanical apparatus used by his famous predecessor, Maupertuis tried to describe the

[126] Maupertuis, *Œuvres*, II, 67.

[127] *ibid.*, 85: "Quoique je respecte infiniment Descartes, & que je croie, comme lui, que le foetus est formé du mêlange des deux semences, je ne puis croire que personne soit satisfait de l'explication qu'il en donne, ni qu'on puisse expliquer par une méchanique intelligible comment un animal est formé du mêlange de deux liqueurs. Mais quoique la maniere dont ce prodige se fait demeure cachée pour nous, je ne l'en crois pas moins certain."

molecular reactions occurring within the womb by means of applying the principle of attraction to embryology. Buffon, to whom the notion of preexistence seemed both practically and philosophically absurd, had shared the dissatisfaction of the *Vénus physique* with the dominant approach to the mystery of generation. Needham's investigations presently furnished Buffon with what he regarded as an empirical basis for his own theories. The first volumes of the *Histoire naturelle* contained an elaborate doctrine of generation which, scrupulously avoiding any finalistic allusions, was founded solely on the supposed mechanical properties of the organic molecules supplied by both parents in the reproductive act. Diderot drew from such speculations, in which he early became interested, the obvious conclusions favorable to a naturalistic interpretation of the phenomenon of life. A statement of epigenesis, with Buffonian echoes, had its place in the *Entretien entre D'Alembert et Diderot* as one of the scientific foundations of materialism. With the repudiation of preformationist thought, Diderot freed biological inquiry from theological bonds:

"DIDEROT: . . . allow me to give you the history of one of the foremost geometricians of Europe. What was this marvelous person at first? Nothing.

D'ALEMBERT: What do you mean, nothing! One cannot make anything from nothing.

DID: You are taking things too literally. I mean that before his mother . . . had reached the age of puberty, before [his father] was an adolescent, the molecules destined to form the rudiments of our geometrician were scattered throughout their young and delicate bodily mechanisms. These filtered through with the lymph, circulated with the blood, until finally they entered the reservoir intended for their coalescence, the testicles of his father and mother. There you have that rare seed formed; then, as is commonly believed, it is led by the Fallopian tubes into the matrix . . . behold it growing by degrees and advancing to the state of a foetus. . . .

D'AL: You do not believe, then, in preexistent *germes*?

DID: No.

D'AL: Ah, how that pleases me!

DID: They are contrary to experience and reason: against experience, which would search fruitlessly for these *germes* in the egg and in the majority of animals before a given age; against reason, which teaches us that the divisibility of matter has a limit in nature, although it has none in the understanding, and which is loathe to conceive of a completely formed elephant the size of an atom, and within that atom another elephant all formed, and so on to infinity."[128]

Of still greater significance for us, however, was Buffon's contribution towards the fulfillment of Descartes's broad biological objectives in the *Monde*. In that regard, Voltaire singled out the historical antecedents of his contemporary's hypothetical constructs: "It is deplorable that the academician who allowed himself to be deceived by Needham's erroneous experiments hastened to substitute his organic molecules for the evidence of '*germes*.' He formed a universe. It has already been remarked that the majority of philosophers, following the example of the chimerical Descartes, have wished to resemble God and fabricate a world by means of words."[129] While Buffon did not at any time profess ma-

[128] Diderot, II, 109-110. Epigenesis was indorsed also by D'Holbach's *Système de la nature*, I, 78.

[129] Voltaire, *Œuvres*, XXVII, 220. It was in particular the universe of organic phenomena, as deduced by Buffon and Diderot, that Voltaire would seem to have had in mind. XXVI, 408: "Il y a plusieurs années qu'un Irlandais, jésuite secret, nommé Needham . . . crut s'apercevoir qu'il avait fait naître des anguilles avec de l'infusion de blé ergoté dans des bouteilles. Aussitôt voilà des philosophes qui se persuadent que si un jésuite a fait des anguilles sans germes, on pourra faire de même des hommes. On n'a plus besoin de la main du grand Demiourgos; le

terialism, the fact remains that his theorizing aided appreciably the growth of that ideology among the philosophes. The speculative features of the *Histoire naturelle* found an apt and promising student in Diderot himself.[130] It ensued from Buffon's biology that the distinction between the organic and the inorganic was no more real than it had been for Descartes: "the living and animated, rather than being a metaphysical degree in the scale of existence, is a physical property of matter."[131] Matter being the stuff from which the mechanical laws of nature were forever manufacturing organic forms, Buffon suggested that, instead of the conventional categories of the organic and inorganic, the new distinction be made between "matière vivante" and "matière morte." Diderot took over Buffon's notion and made it the basis of what was to be his own panvitalist system. In the *Interprétation de la nature*, the differentiation of matter into the "living" and the "inert," which Diderot seemed to accept on the authority of Buffon, was the occasion for stressing the essential homogeneity of material substance as such. The hiatus between the animate and inanimate varieties became simply a mechanical function: "Is any difference to be assigned between *matière vivante* and *matière morte* other than its organization and the real or apparent spontaneity of motion? Is not what one calls living matter merely a matter that moves by itself? And what is called dead matter something that is moved by another material object?"[132] Diderot's reflections on Buffonion biology, striv-

maître de la nature n'est plus bon à rien. De la farine grossière produit des anguilles; une farine plus pure produira des singes, des hommes et des ânes. Les germes sont inutiles: tout naîtra de soi-même. On bâtit sur cette expérience prétendue un nouvel univers, comme nous faisions un monde, il y a cent ans, avec la matière subtile, la globuleuse et la cannelée."

130 The notes (since lost) that Diderot, while imprisoned at Vincennes, compiled on Buffon's work were intended to be useful in a subsequent edition of the *Histoire naturelle*. Diderot, xix, 422-423.

131 Buffon, *Œuvres*, i, 434.

132 Diderot, ii, 58-59.

ing to abolish altogether the gap separating the organic from the inorganic, may thus be said to have approximated more closely Descartes's position than had the *Histoire naturelle*. For Diderot endeavored to reduce the dichotomy between the living and the non-living to an ultimate variability in the inherent motions of a homogeneous matter. A principal goal of the *Rêve de D'Alembert* was to demonstrate, by referring their formation to the same general causes, the continuity existing in the universe from the most inanimate to the most highly organized beings. But this continuity, extending from *matière morte* to *matière vivante*, or (what was equivalent) from *sensibilité inerte* to *sensibilité active*, remained fundamentally a continuity of matter in motion.[133] The materialism deduced from the lessons of vitalist biology, in Diderot's *milieu*, was absorbed into the available context of Descartes's mechanistic philosophy of nature.

The transition from the *bête machine* to the *homme machine* must be understood, then, in the light of the Cartesian influence that acted on the philosophes through the medium of the scientific interests of the 1740's. If La Mettrie claimed that man was a machine, his meaning should not be distorted by the usual semantic associations of the term "mechanical." For La Mettrie, as for the materialist school that owed much to him, the organism was a machine capable of sensation, intelligence, and purposeful behavior. It had been formed, and could in some cases be regenerated, by natural causes. La Mettrie's conception of mechanism attested the dynamism inherent in matter itself: "the human body is a machine that rewinds its own springs—the living image of perpetual motion."[134] Although Buffon was careful

[133] *ibid.*, 106, 109-110, 115. In the *Eléments de physiologie*, Diderot remarked: "Je serais tenté de croire que la sensibilité n'est autre chose que le mouvement de la substance animale, son corollaire, car si j'y introduis la torpeur, la cessation de mouvement dans un point, la sensibilité cesse." *ibid.*, IX, 268.

[134] La Mettrie, *Œuvres philosophiques*, III, 12.

not to define *man* as an automaton, his notion of organism, as given in the *Discours sur la nature des animaux,* likewise represented the fusing together of matter and soul, of movement and sensation. It was Condillac who rightly assumed that Descartes's automatist thesis had been to blame for the materialist tendency of Buffon's ideas. Accordingly, his *Traité des animaux* undertook to confute simultaneously Descartes's and Buffon's theories of organism by means of the same argumentation. Condillac's objections showed an awareness of the peculiarly unorthodox side of the beast-machine's career: "It was not enough for Descartes to have attempted to explain the formation and conservation of the universe by the laws of motion exclusively; he had also to reduce animated beings to pure mechanism."[135] The *Traité des animaux* said of Buffon's treatment of animals that he pretended, "like Descartes, to explain mechanically all their actions."[136] But while Cartesian philosophy had overtly denied sensation to the beast-machine, Buffon had made that quality the product of corporeal functions, which, as we have seen, was simply the alternate interpretation of the automatist hypothesis. Condillac went on to complain that the definition of sensation as an "act of motion, occasioned by impulsion or resistence" would, in reality, mean that "the most brute matter would be sensible."[137] Such a critical perception contains the logical link between Descartes and Diderot with respect to naturalistic biology.

It is with this consequence of automatism, as expressed by Buffon, that the point of departure for Diderot's vision of nature becomes all the more palpable. The *Interprétation de la nature* described the organism as "a system of different organic molecules which, by the impulsion of a sensation similar to an obtuse and dull sense of touch that the Creator of matter has in general endowed it with, have combined with

[135] Condillac, *Œuvres philosophiques,* I, 340.
[136] *ibid.,* 341. [137] *ibid.,* 342.

one another until each has encountered the position most suitable to its figure and repose."[138] Diderot's attribution of sensibility to matter, which thus remained intelligible only through the mechanistic medium, found confirmation in still another scientific event of the time. This was Haller's discovery of muscular irritability, a phenomenon already anticipated by La Mettrie's *Homme machine*, and cited in favor of that work's materialism. Condillac took the principle of irritability to be the immediate basis in fact for Buffon's contention that the animal was an automaton endowed with sentiment.[139] Diderot was to become conversant with the naturalistic import of Haller's famous contribution to physiology: "If life remains in the organs severed from the body, where is the soul? What happens to its unity? What happens to its indivisibility?"[140] No valid obstacle was left against drawing the last conclusion from the long history of the *bête machine*, namely, the identification of soul with matter.

A glance at Diderot's reaction to Maupertuis's biological speculation, which in some ways paralleled his own, will help to make plainer the former's consistency with Cartesian precept. Maupertuis had begun with the conviction that "never will the formation of any organized body be explained by the physical properties of matter exclusively; and from Epicurus to Descartes, one has only to read the works of the philosophers who undertook it to be persuaded of this."[141] Since religion, too, ruled out the possibility that "the bodies which we see owe their first origin to the laws of Nature alone, to the properties of matter,"[142] Maupertuis proposed to restrict his study to merely the laws by which the economy of things functions and conserves itself *ex post facto*. His

138 Diderot, II, 49-50. D'Holbach defined the organism as "une machine sensible." *Système de la nature*, I, 177.

139 Condillac, *Œuvres philosophiques*, I, 346n.

140 Diderot, IX, 275.

141 Maupertuis, *Œuvres*, II, 155-156.

142 *ibid.*, 154.

monadological approach, which endowed the particles of matter with faculties of memory, desire, and aversion, was in actuality, as Diderot fully recognized, a means of clarifying vitalist phenomena without having to admit the materialist consequences. Such an aim was basically in keeping with the *Essai de cosmologie*, in which Maupertuis had already tried to prop the crumbling Newtonian teleology with his "principle of minimum action." His biology in turn represented, as Diderot's synopsis of it suggested, a recasting of the traditional doctrine of animal soul: "If there were . . . any peril in the granting of some degree of intelligence to the molecules of matter, this peril would be as great in supposing it in an elephant or monkey as in a grain of sand."[143]

In such a restoration of intelligent causes to nature, Diderot detected a persistent, albeit tenuous, tie between theology and science. He wrote: "Here the philosopher of the Academy of Erlangen makes desperate efforts to avoid all suspicion of atheism; *and it is evident that he advances his* hypothesis . . . only because it seems to him to satisfy the most difficult phenomena without materialism being a result. One must read his work to learn how to reconcile the most daring philosophical ideas with the profoundest respect for religion."[144] Diderot thereupon proceeded at once to discredit and to utilize Maupertuis's doctrine by inferring pantheism from it: if the particles of matter are possessed of memory, desire and aversion, and if they all form together, as the case must be if nature is a unit, a single organism, then the universe has a soul, and this soul may well be God.[145] Of chief interest here is, not so much Diderot's reading of unorthodox opinions into the *Dissertatio inauguralis*, as the underlying reason for his disagreement with its author. Diderot's dispute with Maupertuis coincided, by and large, with La Mettrie's opposition to Leibniz's monads: to wit, that these tended to spiritualize matter, rather than to make

[143] Diderot, II, 46. [144] *ibid.* [145] *ibid.*, 47-48.

spirit material. Diderot found fault with what he regarded as Maupertuis's dualist bias in the definition of matter. By contrast, he himself favored a biological system that would start out with matter in motion, and then set about to deduce, from the corpuscular combinations and arrangements that followed, such secondary phenomena as desire, aversion, memory, and so forth. The ideal of a mechanistic materialism was made paramount in Diderot's criticism of Maupertuis, when he stated that one must suppose in matter "a sensibility a thousand times less than what the Almighty has granted those animals that are closest to being *matière morte.* In consequence of this dull sensibility and of a difference in configuration, each organic molecule would find a single situation that is more appropriate than any other, which it would have sought by an automatic unrest. . . ."[146] The loyalty of Diderot's thinking to the naturalistic biology that had its inception in Descartes is here revealed in the key words: configuration, situation, automatic. Diderot's position was at the heart of the Enlightenment's materialist science: that of Maupertuis, torn between the equally insistent demands of the old teleology and of the new vitalist current, remained on its periphery.

The concept of transformism, which anticipated so brilliantly the future course of biological progress, was an integral part of the ideology of the *côterie holbachique.* But the background and motivation of the evolutionist ideas diversely voiced by La Mettrie, Maupertuis, Diderot, and Buffon has remained, despite long-standing study, in some ways obscure.[147] Scholars have for the most part agreed that the earliest manifestations, around 1750, of this pre-Lamarckian

[146] *ibid.,* 49-50. Stated more plainly by D'Holbach, it was owing to a mechanical disposition that "les molécules primitives & insensibles dont tous les corps sont formés, deviennent sensibles." *Système de la nature,* I, 52.

[147] For a general discussion of the subject, consult Emile Guyénot, *Les Sciences de la vie au XVIIe et XVIIIe siècles. L'idée d'évolution,* Paris, 1941, pp. 337-401.

doctrine were not, strictly speaking, of a piece with its later versions based, more directly, on the cumulative evidence of mutations, comparative anatomy, natural selection, and other empirically verifiable data. The transformism—"unscientific" by modern standards—of Diderot and his immediate associates had a different inspiration. It came to the fore, on the whole, as a theoretical "by-product" of the cosmogonical, geological, and biological speculations and problems that made up the trend of naturalism which, in the foregoing pages, has been traced from Descartes down to the philosophes.

The materialists of the Enlightenment, taking their cue from Cartesian science, claimed that the cosmos and its specific contents both organic and inorganic were the intelligible outcome of mechanical process. Such an assertion clashed sharply with theological authority, which dominated scientific thought to the extent of reserving to itself the right to pronounce on (among other topics) the manner of the world's creation, the origin of species, the providential ends served by external patterns, man's entry upon the scene and the destiny peculiar to him, and related subjects. The Church in France (like many Protestant sects elsewhere) laid down dogmatically, against all naturalistic explanations, the description of what had happened given in *Genesis*, and eventually rested the case for finalistic biology on the presumed regularity of generation by means of specially contrived *germes*. The task of showing, to the contrary, how organic forms could have developed uniquely from the properties of matter and the laws of motion became, consequently, synonymous with atheistic materialism. The transformistic theme put forth by certain philosophes around the mid-century was, in the first instance, a hypothesis meant to illustrate, even in the absence of concrete proofs, how the various species of vegetable and animal life might have resulted from purely physical causes. As such, it represented the latter-day completion of the Cartesian *Monde*. Had Descartes himself, during the long years of groping with the problem, discovered and dared to use the same general formula, it might conceivably have brought his system of nature to the contemplated goal.

The genesis of the Cartesian world had incurred, necessarily, a dynamic notion of *res extensa.* Matter was believed capable, by its mechanical attributes, of developing through successive states until it had *transformed* an original chaos into the present scheme of things. The *Principes de la philosophie* expressed clearly this conception:

"Besides, it is of little importance in what manner I suppose matter to have been disposed in the beginning, since its subsequent rearrangement must be effected according to the laws of nature. And it is hardly possible to imagine any disposition of it from which one could not prove that matter, by these laws, must continually change, until it has finally composed a world entirely similar to the actual one (although the deduction would perhaps be longer from one supposition than from another). For these laws being the cause that matter must successively take on all the forms of which it is capable, if one considers in order all these forms, it will be possible in the end to arrive at that which is now to be found in the world."[148]

Leibniz was among those who apprehended that deterministic naturalism was implicit in this passage from the *Principes*; that a serious threat to natural theology ensued from Descartes's developmental view of matter in motion.[149] Whether Leibniz's appraisal of the issue was accurate or inaccurate

[148] Descartes, IX, 126.

[149] It was said of Descartes's summary of physical process in the *Principes*: "Son Dieu fait tout ce qui est faisable et passe, suivant un ordre necessaire et fatal, par toutes les combinaisons possibles: mais à cela il suffisait la seule necessité de la matiere, ou plustost son Dieu n'est rien que cette necessité ou ce principe de la necessité agissant dans la matiere comme il peut." Leibniz, *Die philosophischen Schriften*, herausgegeben von C. J. Gerhardt, Berlin, 1875-1899, IV, 299, and *passim.*

(students of philosophy have argued both sides[150]) is not particularly relevant here, since its significance for us consists mainly in the bright light it projects along the historical path taken by Descartes's physics. As seen in Leibniz's remarks, the broad theoretical frame of what was to be the philosophes' evolutionary materialism was already present in the *Principes*. It devolved upon the "unavowed" heirs of Cartesian science to attempt to show in detail how matter, in order to form the species extant, had passed through numerous possible states. La Mettrie inserted this consideration amid the somewhat disconnected insights which, in the *Système d'Epicure*, made up his evolutionist outlook: "Through what infinite number of combinations has matter had to pass, before attaining that unique combination from which a perfect animal could have resulted!"[151] Likewise Diderot's discussion of transformism, rich in Lucretian reminiscences, conjectured about the probable preliminary states which matter, developing by inherent laws, would have to undergo before it struck certain arrangements "whose mechanism implied no contradiction, and which could subsist by themselves and reproduce."[152] Once again, Descartes had indicated the scientific uses of a notion that went back to classical Epicureanism.

The vitalist facts made available during the 1740's were the immediate motivation for renewing in earnest Descartes's as yet unsuccessful enterprise. "When I have seen inert matter passing to the state of sensibility," exclaimed Diderot, "nothing else ought to astonish me. . . . You have two great phenomena: the changing of matter from an inert to a sensible condition, and spontaneous generation. Let these suffice.

150 Cf., for example, Jean Laporte, *Le Rationalisme de Descartes*, Paris, 1945, pp. 343-361. However, the author's treatment of the question, conforming with the idealistic exegesis of Cartesian thought, attempts an internal reconciliation of Descartes's non-finalistic idea of matter with theological orthodoxy.

151 La Mettrie, *Œuvres philosophiques*, III, 222.

152 Diderot, *Lettre sur les aveugles*, p. 42.

Draw the appropriate consequences. . . ."[153] A major consequence was evolutionary materialism. With the obliterating of the gap between the organic and inorganic, and the spontaneous birth of organisms such as Needham's animalcules, there remained no further barrier against supposing that matter in motion (or fermentation) had engendered the world and its contents. What was needed was merely a hypothesis that could present the process as feasible. In the search for it, the transformistic hypothesis finally imposed itself by its own merits.

As noted earlier, the view of organic variability was already implicit in a cosmogonic system that, like Descartes's *Monde*, set out to describe the transformations of the cosmos as a whole from chaos to its actual order. Alterations in the species of living things could, by the logic of the situation, be expected to correspond in some manner to the normal vicissitudes of the world in which they occurred. That Nature might thus be capable of new productions from time to time, despite periods of inactivity such as the present one, was hinted in the version of organic origins given by Descartes's disciple, Régis. Conceding that motion had metaphysically been communicated to matter by God, Régis nonetheless maintained that as a result of this motion matter "was divided into parts, several of which had precisely the size and figure necessary to make them assume the order and arrangement in which the form of the two first beasts of each species was to consist. There is even reason to believe that the material particles disposed themselves in that fashion by following the same laws of motion that are observed today; and that if we do not witness the appearance of new species of animals, this is owing to the fact, not that the laws of nature have changed, but that matter now has sensible qualities which it did not possess when the original animals were formed."[154]

[153] Diderot, II, 134.
[154] Régis, *Système de philosophie*, II, 508.

Fontenelle gave his recapitulation of the issue: "When the earth was constituted, being full of lively and active atoms, impregnated with the same subtle matter of which the stars had just been composed . . . it could have been fecund enough to give rise of itself to all the different species of animals."[155] To which Fontenelle, whose Cartesianism was not always consistent, stated two reservations favorable to finalism: first, how could animal life have subsisted under the conditions imagined necessary for its origination; and second, how come that nature now generates regularly by means of *germes* if it once produced organisms spontaneously? The solution of just such difficulties was to stress the point that plants and animals need not have been in the beginning what they are now; that, together with their reproductive mechanisms, they might well have varied in physical relation to the successive stages of the globe's cosmogonical and geological development.

Two manuscript works of the first half of the eighteenth century, the *Recherches curieuses* and the *Dissertation sur la formation du monde*, attest concretely the many-faceted trend of scientific speculation that concerns us here. Both treatises, as shown in a previous chapter, were continuations of Descartes's project of natural philosophy, and even recounted their manufacture of a world from matter in motion against the specific background of the vortex physics. The inferences that this led to, in either case, regarding the problem of organic origins may be considered as a prelude to the philosophes' evolutionary materialism.

The *Recherches curieuses* attributed the first emergence of organized beings to the interplay of certain purely chemico-mechanical causes.[156] It attempted, however, to correlate this process with the fluctuations of a much broader physical environment. Thus, the decisive event occurred when the earth

155 Fontenelle, *Œuvres*, III, 230.
156 *Recherches curieuses*, pp. 363-378, 400, and *passim*.

found itself at a given moment in the precise proximity to the sun required for the generation and sustenance of living things.[157] Life appeared initially near the equator, and then, as our planet continued to warm up, different varieties of it, depending on local conditions, sprang up spontaneously ever closer to the poles, until in the end "the earth was covered with herbs, plants, trees, and furnished with animals of all kinds."[158] The anonymous writer's theory, it is true, is not yet authentically transformistic; for although new species are said to come into existence at various epochs of the globe's natural history, it is assumed that these were all independent productions, not derived one from the other. Nonetheless, the envisagement of living types from a standpoint of temporal change has broken, in a definitive sense, with the theological tradition of their simultaneous Creation *in toto* and their presumed invariability ever since. Ideologically, transformism is justified. In addition, the *Recherches curieuses* referred dissimilarities among animal species, and even among the races of mankind, to climatic or physiographic factors[159]—a theme that Buffon was later to elaborate with considerable success before fitting it into an evolutionary doctrine. Elsewhere, anthropological information was utilized to deny any essential distinction between man and beast, and to illustrate how the one could result from the other. Certain savage tribes of the Dutch East Indies were taken as examples of an intermediary stage of rationality between the higher mammals and civilized man.[160] The entire subject of the origins and growth of intelligence was treated naturalistically. The "spiritual" faculties, or rather their variations in civilized man, the savage and the animal kingdom, were explained in terms of a composite of causes remaining strictly on the material level, such as physiological structure, physical heredity, environmental experience, climate, and so on.

[157] *ibid.*, 388. [158] *ibid.*, 392. [159] *ibid.*, 393ff.
[160] *ibid.*, 447-451, 454-455.

Much nearer in time to the philosophes, the *Dissertation sur la formation du monde* reflected the increasing interest in biology to be met with already during the 1730's. Also, as might be expected, its materialism was bolder and more coherent. Taking up the question of organic origins where Descartes's *Monde* had left it, the unknown author started with a mechanical account of the genesis of all animate beings. The pieces destined to form these latter having been present in "universal substance," it was held that "the laws of ponderability and the property of extension could not have failed to assemble and unite a large number of corpuscles of different kinds, after the formation of the globes, and to make from these distinct bodies, some composed of homogeneous, others of heterogeneous, elements. The former became organisms; the latter must have perished, at least with respect to their forms."[161] It is assumed that the earth was once in a state suitable for the spontaneous generation of life; that, as certain physical requirements were realized at various times and places on its surface, the corresponding species of plants and animals, some of which have since vanished, appeared by natural causes.[162] The fundamental requirement was, of course, the right temperature; life must, therefore, have come last to the polar regions, which long lacked the degree of heat needed for "fermentation."[163] The whole process of peopling the earth with organic creatures "took probably a great number of centuries."[164] But the process, while interrupted within the range of man's memory, could recommence. The most productive period in the past came soon after the retiring of the global seas, when humidity was at its highest point; the "drying out" of things has since arrested new productions. However, "in case we could

161 *Dissertation sur la formation du monde*, p. 127.

162 *ibid.*, 132ff., and *passim.*

163 *ibid.*, 158-159; "Les productions se sont faites à raison des climats" (p. 168).

164 *ibid.*, 157.

restore the slimelike quality that the earth's surface had in the epoch when organisms were formed, who knows if we would not see the appearance of a new world of species!"[165] The multitude of extant species is explained by the countless possible arrangements of the particles of matter; and Nature has no more difficulty in making an elephant than a frog, provided only that matter finds itself in the appropriate combinations.[166] The *Formation du monde*, like the *Recherches curieuses* with which it obviously has much in common, does not advance a true transformism, but adopts a theory of "isolated creations." But of special interest is the fact that its author discusses, in 1738, the prototype doctrine as something advocated by several of his contemporaries who, presumably, were struggling with the same general problem of a naturalistic biology.[167] Indeed, the available evidence shows that Maillet's *Telliamed*, which first made extensive use of the prototype notion, was written at just about that time.

The *Formation du monde* offers, not only a developmental approach to organic patterns, but draws conclusions from its scientific speculations that come extremely close to the evolutionary materialism of Diderot's group. Matter in motion having engendered all things from minerals up to the highest animals, there can be no basic breach between the animate and the inanimate; instead it is affirmed, in a manner foreshadowing Diderot's own panvitalist views, that "except for the world's outer crust, everything has life, for everything has movement: movement is the essential property of life."[168] Along with this, the phenomena of sensibility and

165 *ibid.*, 165. 166 *ibid.*, 161, 165-166. 167 *ibid.*, 128-129.

168 *ibid.*, 131. One finds already in the *Formation du monde* the concept, soon to be associated with Buffon and Diderot, that vital principle is generally present in matter as such, in either an active or passive state: "Ce seroit agir contre l'expérience, que de refuser un genre de vie aux pierres: leur accroîssement suppose le mouvement; et le mouvement est le caractère essentiel de la vie. Cette vie, pour être passive, n'est pas moins une vie. . . ." (p. 147). See also, p. 149.

intelligence are regarded as particular aspects of the shaping of the natural order as a whole: "The first men on earth reasoned no more than an oyster."[169] Nothing, the reader is told, has confused the question of organic origins so much as the decision to consider motion and sensibility as entities apart from matter, "whereas they are simply modifications, or modes of being, of universal substance."[170] With the definition of sentiment as the effect of the animal economy, the beast-machine finds a proper place in this complex of evolutionary materialism.[171] And by a verbal twist, human intelligence is finally included in the picture: "The ideas, like the sensations, of all animals are nothing but the result of our organization."[172] In short, the *Formation du monde* not only exploited the full measure of naturalism implicit in Descartes's *Monde* and *Principes*, but was a good instance of the historical linkage between Cartesian science and the ideology of Diderot's school.

Another such instance was *Telliamed*. We have noted in an earlier chapter how Benoît de Maillet sought, against the background of *tourbillon* physics, to describe the globe's formation from matter in motion. It was in relation to the successive phases of that long geological event that he proposed, likewise, to elucidate the emergence of the different species. In this case, however, a new hypothesis was tried. An organic prototype was supposed, by its diversifications, to have yielded all other living forms. This process of transformism, going (despite a great many omissions) from prehistoric sea-life to modern man, was accomplished in response to the earth's ever-changing environment. *Telliamed* was first published in 1748; within a few years, Diderot, Maupertuis,

169 *ibid.*, 173, and ff. 170 *ibid.*, 177.

171 *ibid.*: "Le sentiment . . . est l'effet de choses qui elles-mêmes n'ont point de sentiment." Descartes was wrong, we are told, to deny sensibility to matter in an absolute sense; for while this might be true of individual corpuscles, it is not true of the assembled machine.

172 *ibid.*, 209.

and La Mettrie had seized upon its thesis, however fantastically presented by Maillet. Without insisting that the evolutionary ascent had begun from aquatic creatures, nor that it would end with man, the materialists took firm possession of the crucial idea that organic types were subject to transformation in keeping with more general cosmic developments. Transformism came up among the philosophes as an integral part of the Cartesian plan to reconstruct theoretically a world from matter in motion. As such, it was a decisive answer to those who, to grant finalistic theology a foothold in the sciences of nature, had assiduously claimed that no generation was comprehensible without deliberately contrived *germes*:

"D'ALEMBERT: But without preexistent seeds, the initial generation of animals cannot be conceived.

DIDEROT: If the question of whether the egg or the chicken came first embarrasses you, it is because you imagine that animals have been originally what they are now. . . . It is no more known what they have been than what they will become. The imperceptible worm that moves in the mud is perhaps on its way upwards to the state of a large animal. The enormous animal . . . moving downwards to the state of a worm is perhaps a particular momentary production of this planet."[173]

In the aggregate of ideas usually denoted as pre-Lamarckian transformism, there were of course a number of indispensable ingredients, both scientific and philosophical in character, the examination of which does not lie within the confines of the present study. Our purpose is to point out, primarily, the hitherto neglected role and value of the specifically Cartesian contribution to the rise, around 1750 in France, of the evolutionist outlook.

It is known that the chain-of-being concept was an essen-

[173] Diderot, II, 110.

tial component of the intellectual attitude in question. Notwithstanding this, the postulate of an unbroken gradation in nature, taken alone, was insufficient to support the theorizing of Diderot and his colleagues. Logically it opposed neither traditional belief in the simultaneous Creation of all species, nor theological suppositions about their fixity ever since. Moreover, it could not have furnished the materialist setting in which transformistic thought first occurred. Indeed one could, during the Enlightenment, combine an unreserved acceptance of the *chain-of-being* with the latest knowledge in botany and zoology without, for that reason, feeling obliged to embrace transformism. The most outstanding among numerous instances of this was, perhaps, Charles Bonnet. Before the Leibnizian law of continuity could be utilized ideologically by the philosophes, it had to be reinterpreted. Viewed through the prismatic implications of Descartes's physics, it came to represent, not so much an immutable hypostatic series, as a temporal sequence of organic forms realized by matter in motion.

Glancing at the fund of scientific data available around 1750, its inadequacy for the framing of an evolutionary hypothesis likewise becomes fairly apparent. The soundest empirical basis for pre-Lamarckian doctrine was to consist in the testimony of mutations. At the mid-century, however, not only was such evidence still quite scant, but the most influential scientists of Europe were engaged in circumscribing its applications. Linnaeus, for example, who from religious scruples dogmatically upheld the constancy of species, classed mutationist phenomena with hybridizations; and his great reputation, founded on such widely-read works as the *Systema naturae* (1735) and the *Philosophia botanica* (1750), retarded for a long time the advance of transformism among professional and amateur naturalists alike. It is true that Maupertuis, investigating problems of heredity, had discussed mutations at some length in the *Vénus physique* of 1745; but his reflections, far from being generalized into an evolutionary conception, were restricted to explaining merely variations within each species, not variability among the species themselves. Buffon's statement of transformism, for the most part of considerably later date, was actually to make use of proofs drawn from mutations, comparative anatomy, paleontology, and the geographical distribution of extant species; but these same materials were hardly available, around 1750, in either the kind or quantity required for a cogent inference of evolutionism. The transformistic hypothesis of de Maillet, Diderot, and La Mettrie, accompanied by remarkable insights into natural selection and preadaptation, would seem largely to have been the type of semi-philosophical speculation which, in the history of ideas, often closely anticipates an entire trend of scientific discovery and demonstration.

As such, the earliest expression of organic evolution among the philosophes came as the final theoretical outgrowth of Descartes's *Monde*. Maupertuis, referring in 1751 to the current vogue of transformism in some circles, revealed its intimate ties with the Cartesian-Buffonian project of a mechanistic cosmogony:

"This is not the place to recount the changes that seem to have taken place in our globe, nor the causes that could have produced them. . . . But it is possible to start from the fact that all forms of matter on the earth's surface, as we have every reason to believe, were once in a fluid state, whether dissolved in water or molten by fire. Now, in this condition, the materials of our globe were in the same state as those liquids in which the elements destined to form animals float about; and metals, minerals, precious stones were much easier to produce than the least organized insect. The least active elements of matter have brought into being metals and marble; the most active, animals and men. The whole difference between these productions is that some still go on owing to the fluidity of the matter in which their component parts are present, while the hardening of the substance in which the ingredients of the others occur does not allow of new creations. But it would not be impossible, should our planet find itself, as a result of a deluge or conflagration, in one of the states of which we have spoken, that new unions of the elements, new animals, new plants, or rather, completely new things, would make their appearance. It is thus that with a single principle all the objects could be explained about whose production we are today wholly ignorant. In the former fluid condition of matter, each element could have placed itself in a manner appropriate for shaping those bodies in which one is no longer able to discern any vestige of their formation."[174]

Similarly, Diderot conceived of evolutionary process in relation to the vicissitudes of the cosmos, and as the organic counterpart of that rise and fall of worlds described and popularized by Cartesian physics:

"DIDEROT: Would you permit me to anticipate time by several thousand years?

D'ALEMBERT: Why not? Time is nothing for nature.

DID: You consent, then, that I extinguish our sun?

D'AL: All the more willingly since it is not the first time that the sun has been extinguished.

DID: With the sun cold, what will happen? The plants will perish, the animals will die, and the earth will be solitary and silent. Rekindle that star and you reestablish at the same time the necessary cause for an infinite series of new generations, among which I dare not assert whether, after the passing of centuries, our plants and our animals will or will not reproduce themselves."[175]

[174] Maupertuis, *Œuvres*, II, *152-154.
[175] Diderot, II, 111.

The *Système de la nature*, where D'Holbach brought together the diverse strains of materialist ideology in his period, made known the same nexus between Cartesian science and eighteenth-century transformism. D'Holbach started with a naturalistic cosmogony obviously inspired by the *Monde*: "Nature by its combinations engenders suns, which place themselves at the centers of so many solar systems; it produces planets that gravitate by their own essence and perform their revolutions around those suns. Little by little, movement alters both suns and planets, and will perhaps someday disperse the pieces with which it has composed these marvelous masses. . . ." The *Système de la nature* evokes, thereupon, the Cartesian principle of constancy of motion as the cause of all change in the universe. This in turn leads, by a bold but logical step, to a standpoint of biological evolution: "It is, therefore, the continual motion inherent to matter that alters and destroys all organic beings, that deprives them at each moment of some of their properties in order to replace these with others. It is motion which, in thus varying their present natures, changes also their orders, directions, tendencies, and the laws governing their modes of existence and activity."[176] Such was the outcome, in the Enlightenment, of Descartes's century-old proposal to derive, from the successive possible combinations of moving matter, "all things in the world, both general and particular."

The comments made by contemporary critics of scientific naturalism tend to confirm the various themes of ideological continuity mentioned in this chapter. The first theme had to do with the metamorphosis of the beast-machine into the man-machine. Gautier d'Agoty, dealing in 1754 with that point, observed: "Descartes and several philosophers of antiquity have made animals into machines; from which it has been concluded that man himself, being composed of blood, veins, viscera, nerves, and muscles, is also a mechanical being.

[176] D'Holbach, *Système de la nature*, I, 42-43.

I admit, if it could be proved that animals act only mechanically, then by comparing human actions with those of beasts one could demonstrate that our thought is merely a modulation of matter . . . in a word that all our expressions are simply material motions."[177] D'Agoty stated that Cartesians and Epicureans, although beginning with different definitions of matter, had reached the same general opinion concerning the materiality of the animal-soul.[178] He pursued: "I ask . . . those thinkers who understand me whether it would not be possible to say as much of man, if one granted the materialistic Cartesians what they affirm about beasts?"[179] Soon after, the abbé Dufour reviewed the same question and gave his appraisal of the logical consequences of Descartes's mechanistic physiology: "Everything that takes place with regard to human beings and animals ought not, since it is explicable by the impressions of a more or less subtle matter and the general laws of motion, to be explained in any other fashion. For it is absurd to multiply substances without need, and to clarify by means of an unintelligible principle what can be accounted for by means of a clear, simple, and tangible one."[180] Once a uniformly iatrophysical biology was believed capable of elucidating all organic functions, no amount of metaphysical argument could succeed in proving that "man is animated by an immaterial substance that can neither be seen, nor known, nor felt."

But the automatist thesis attributed, not only the functions, but the formation too of organized beings to mechanical causes—an attitude that was to be extremely advantageous to materialist thought. Denesle said about it in 1754: "The partisans of animal-automatism presuppose gratuitously what no one concedes to them, namely, that organization and vegetation as such are able to impart to matter an

[177] Gautier d'Agoty, *Observations sur l'histoire naturelle*, Partie XI (1754), pp. 3-4.

[178] *ibid.*, 5.

[179] *ibid.*, 6.

[180] Dufour, *l'Ame ou le sisteme des Matérialistes*, pp. 22-23.

invincible inclination towards one form as against another. From this system to that of the Materialists is but a step, and the induction of one from the other is entirely natural."[181] The abbé Gauchat, examining explicitly the link between Cartesian cosmogony and eighteenth-century naturalism, inveighed against Maillet's *Telliamed* and Buffon's *Histoire naturelle* as specific instances of such an intellectual affinity.[182] Another dissenter, alluding to the *Interprétation de la nature*, noted that Diderot's tranformist doctrine had found its incentive in the still persistent aims of Descartes's *Monde*: "You know that one of the great embarrassments of those fabricators of worlds, who want to owe everything to matter, is the formation of men and animals. When, how, and with what have they been constructed? Quite recently, these mysteries have been unveiled, and the origins of organic beings discovered."[183] In doing that, Diderot and his group had extracted the maximum degree of materialist philosophy from the biological phase of their Cartesian heritage.

181 Denesle, *Examen du matérialisme*, I, 385.

182 Gauchat, *Lettres critiques*, xv, 215ff. and 281ff.

183 Tiphaigne de la Roche, *Bigarrures philosophiques*, Amsterdam & Leipsik, 1759, II, 17-18.

CHAPTER V

SUMMATION

La nature a donc été avant l'art, [qui] *s'est formé sur ses traces.*

—La Mettrie, *Œuvres,* III, 225.

CHAPTER V

SUMMATION

From the network of evidence presented in the foregoing chapters, certain conclusions are indicated. The naturalism of Diderot, with the collaboration it received in the varied efforts of La Mettrie, Buffon, D'Holbach, and even Maupertuis, was in essence motivated by the desire to solve definitively the general problem of a science of nature. Just what such a natural science was to be; where its first principles were to be sought; how far its powers of explanation might apply to the physical universe—had all been questions for which, under the reigning theological tradition, no frank or unbiased answer could be given. A solution was forthcoming, as it happened, only through an ideology which by its own *a priori* assumptions would attempt to discredit and displace that theological tradition and the affiliated dogmas that impeded the progress of rational inquiry. The philosophes' evolutionary materialism was such an ideology.

By defining the sense of their doctrine and by tracing its historical growth, we have seen that the members of Diderot's group were, in the main, heirs to certain methods and aims of Descartes's natural philosophy. The factors that proved to be indispensable to their materialism may be summarized as follows: (1) the decision to explain all phenomena, from the broadest cosmic outlines to the most intricate organic details, by ultimate reference to determinable laws of matter in motion; (2) the elimination of all vestiges of finalism, providential designs, plastic natures, hylozoistic schemes, and other concepts adaptable to apologetic use, from the program of the investigator of nature; (3) the demand that the scientist give an account not merely of the operations, but of the origins as well, of the external order. This special synthesis of desiderata and criteria had of course been as typical of Descartes as it came in time to be of Diderot and his associates. The sources and meanings alike of scientific naturalism in the eighteenth century are thus most intimately, although by no means exclusively, discoverable in the complex of Cartesian thought.

True, the ideas of Diderot, La Mettrie, Buffon, and D'Holbach made up, when taken comprehensively, a rather multiform, shapeless, and sometimes self-contradictory doctrine. Its seeming inconsistencies or ambiguities, however, were owing less to any serious confusions or disagreements on the part of the philosophes themselves, than to extraneous features introduced by them for reasons of polemical mystification. This resulted in the inability of their critics always to distinguish the essential from the incidental aspects of their argument. Concerning its philosophical background and affinities, moreover, the names of Epicurus, Lucretius, Gassendi, Hobbes, Spinoza, and Locke were most often invoked on all sides. Undoubtedly all those thinkers made important contributions, and special studies, exploring beyond the limits of the present work, are required to estimate accurately the kind and measure of influence exercised by each. Nonetheless, it may be presumed that not one of them had expressed, with the pertinence and effectiveness of Descartes, the ideological composite that was to comprise the very core of Diderot's attitudes. Consequently, we must be cautious not to take quite literally the philosophes' own statements of intellectual descent. If each of the above-mentioned names is considered separately, the inappropriateness of making any one the crucial forerunner of scientific naturalism will, perhaps, become more apparent. At the same time, it will be possible to characterize briefly the positive role played by each of those philosophers in the rise and diversification of the current of thought that culminated in the *côterie holbachique.*

Gassendi, the official reviver of Epicureanism, was most likely Descartes's closest rival and partner in the fostering of eighteenth-century materialism. The former's atomistic teaching had early merged with and fortified the Cartesian corpuscular theory; together the two conceptions lent complete support to the pretension of explaining all natural phenomena in terms of bodies in motion. In the sphere of physiology, the tendency of Epicurean tradition to define the soul in relation to certain highly refined atomic movements was easily assimilated to Descartes's automatist hypothesis and to the man-machine resulting therefrom. In addition, the achievement of Gassendi was chiefly responsible for the appearance in France, between 1650 and 1750, of an "Epicurean school." Although it left no profound works in either philosophy or science, libertinism did create among the cultivated classes an atmosphere of freethinking in which the radical implications of Cartesian physics found conditions favorable to their development. In this sense, Gassendism and Cartesianism may be said to have complemented one another in the preparation of the Enlightenment's materialist phase.

But thereafter the parallel ceases, and Descartes comes vividly to the fore. Gassendi, it is recalled, did not extend his ideal of atomistic mechanism to the biological domain as consistently and thoroughly as did Descartes; instead, he brought final causes into the clarification of organic patterns, and failed to appreciate the maximum scientific value of the animal-automaton. Corresponding with this, he ruled out the possibility of accounting intelligibly for the formation of the actual cosmos from a primordial chaos by successive mechanical changes. Gassendi was prone to accept dogmatically, as were all defenders of orthodoxy in the age of *lumières*, the Mosaic "physics." The simple removal of Gassendi's theological restrictions from the remainder of his system would not, moreover, have yielded a natural philosophy equivalent to the Cartesian, since the fusion of theology and physics in the *Syntagma* had been such as to furnish no analogues to those traits of the *Monde* and *bête machine* that logically prompted scientific naturalism. The affinity of Gassendism with the position of Diderot and his colleagues must, for these reasons, be sharply curtailed; its influence should be gauged in terms of the deeper and more far-reaching stream of Cartesian materialism to which, on various commonly held opinions, it was historically subordinated. Cyrano de Bergerac's *Estats et Empires du Soleil* perhaps set the precedent in this respect. All in all, Gassendi's confused—and confusing—prevarication over the equally sterile notions for natural science of atoms-and-chance and providential finality had fallen wide of the peculiar conception of Nature that was to underlie the philosophes' thinking.

Hobbes presents quite another problem. His natural philosophy, like that of Descartes and Gassendi, was fundamentally corpuscular and attributed all objective phenomena to the effects of bodies in motion. But unlike Gassendi, Hobbes wished to dispense with final causes and their theological adaptations, following in that regard, like Descartes, the example of Bacon. Samuel Parker had doubtless been right, in the *Disputationes de Deo et divina providentia* of 1678, to class together and confute the presumed materialism of both Descartes and Hobbes. The latter would seem, so far forth, to be doctrinally in agreement with the members of Diderot's circle.[1] However, two decisive considerations serve to relegate the influence of the English atheist to a secondary, and even peripheral, place. Hobbes made no attempt to deduce from physical causation the emergence of the universe out of an initial chaos, but was satisfied to adhere, on that basic point, to the view of *Genesis*; there was no de-

[1] On Hobbes's ties with Diderot, see Leland James Thielemann, *The Tradition of Hobbes in Eighteenth-century France*, 1950, pp. 227-237, MS in Columbia University Library.

velopmental dimension to his mechanistic interpretation of nature. The best case made by an eighteenth-century critic for the impression left by Hobbesian thought on the French materialists is, to our knowledge, Pluquet's *Examen du fatalisme* (1757). This work analyzed the two principal sources of deterministic attitudes in the period as Descartes and Spinoza on the one hand, and Bacon and Hobbes on the other. Even so, Pluquet had to lessen the responsibility of Hobbes with the admission that "he believed the world had always been, and would always be, what it is. He therefore did not examine either how the heavens had been formed, nor how animals had been produced, but only the laws observed by nature in the eternal revolution of phenomena."[2] Further on, he quoted Hobbes as having declined to speculate about the origins of the world, and as having affirmed a credence on those topics in Holy Scripture, miracles, national customs, and the like.[3] It is apparent that Hobbes, whatever his private persuasion, fell short, so far as natural philosophy was concerned, of the special synthesis that eighteenth-century materialists were to find so conveniently present in Descartes. If this alone were not enough to limit the possible contribution of the Englishman, another invincible reason may be added. Despite Hobbes's considerable impact on the Enlightenment in the realms of political-social theory and of epistemology, the fact remains that he did not inspire, during the period in France from 1650 to 1750, a trend in scientific inquiry in any sense comparable to what Cartesianism has been shown to have produced.

Spinoza, however, participated more significantly in the genesis and dissemination of naturalistic ideas in the first half of the eighteenth century. Scholars have long been conscious of the prevalence of certain Spinozistic elements in the philosophes' ideology, even if it has not as yet been easy to

[2] Pluquet, *Examen du fatalisme*, I, 344.
[3] *ibid.*, 348-349.

ascertain in what manner and to what degree this was so. The difficulty, which can be settled only by a careful examination of the action of Spinozism in the Enlightenment, is owing mainly to the fact that the philosophes, whether of the materialist or deist camp, possessed no definite comprehension of Spinoza's own intent. As it suited their purpose, they cited the author of the *Ethica* rather indiscriminately and, one might say, promiscuously in favor of a variety of opinions that had in common an opposition to the traditional theology. Facilitating this practice was obviously Spinoza's unique, and perhaps unmerited, reputation at the time as the atheist *par excellence*. While interest in Spinoza arose from several different motives, the somewhat misconceived popularity of his pantheistic doctrine ought, by virtue of the data already adduced, to be ascribed to its usefulness in providing a modicum of metaphysical setting for the naturalism derived from the fund of Cartesian science. It has been perceived how often the conclusions drawn by the philosophes from Descartes's physics and biology found utterance in conjunction with a vague pantheism. Bayle had already made known the salient congruities of Spinoza's system, as propounded in an article of the *Dictionnaire historique et critique* devoted to that subject, with a suitably unorthodox version of Cartesianism. Reduced to a convenient formula, the former proved to be a transitional stage in the gradual transformation of Descartes's philosophy of nature into scientific naturalism. Saint-Hyacinthe, La Mettrie, Diderot, D'Holbach, to mention a few, all bore witness to such an employment of Spinoza's ideas. In this connection, La Mettrie could in 1751 make a rather striking identification, in passing, of the man-machine thesis with the meaning of the *Ethica's* "human bondage": "According to Spinoza . . . man is a veritable automaton, a machine subject to the most rigorous necessity, led by an impetuous fatalism, like a vessel by the current of the waters. The author of *L'Homme machine* seems to have written his book expressly in defense of this sad truth."[4] In the *Rêve de D'Alembert*, likewise, pantheistic overtones were present, albeit as an incidental feature of Diderot's exposition.[5] Several concepts understood in a sense cruder than that intended by Spinoza had, from the first, corroborated neatly the major tenets of materialist science. These were: the rejection of the conventional teleology; the granting of autonomy and self-determining powers to matter; the denial of a supernatural order distinct from the natural; the reduction of moral laws to the physical; and so on.

Notwithstanding such themes of continuity from Spinoza to the philosophes, his influence on them, which was general rather than specific in character, found itself circumscribed by two conditions. First, not only did the philosophes have scant appreciation, and no more than a superficial grasp, of pantheism—or of any metaphysics for that matter—but one may surmise that they did not have a very precise knowledge of Spinoza's texts.[6] Besides this, inasmuch as their naturalism had its central motivation and value in the actual progress of physical inquiry, or, more explicitly, in the exploitation of the rich theoretical possibilities of Descartes's *Monde* and *L'Homme*, the teaching of Spinoza could only have remained in an indirect and non-essential, even if otherwise useful, relationship to it. Because of this situation, those who took the trouble, such as Condillac, to refute Spinozism in its technical acceptation, hoping thereby to overthrow the "atheistic" materialism of the Enlightenment, happened in reality to be aiming at a sort of decoy. Such refutations, for whatever they were worth, had no discernible effect on the activities of the materialist school and hardly ever evoked

[4] La Mettrie, *Œuvres philosophiques*, I, 269. The remark was first interpolated in the text of the *Abrégé des Systèmes* in the 1751 edition of La Mettrie's works.

[5] Diderot, II, 142.

[6] As late as 1771, Voltaire's "Questions sur l'Encyclopédie" spoke of "la foule de ceux qui crient: Spinosa! Spinosa! et qui ne l'ont jamais lu." *Dictionnaire philosophique*, I, 391.

a response. For, unknown to the proponents of theological orthodoxy, the philosophes were interested in Spinoza's system not for its own sake as a coherent metaphysics, but as a catalyst in the rise of scientific naturalism from the Cartesian background.

As regards the ties between Locke and Diderot's group, the findings of the present study tend to modify somewhat the opinion most widely held by scholars. This has been that English sensationalism, introduced into France during the 1730's and allying itself there with the traditions of empiricism and skepticism already typified by Gassendi, Bayle, and their followers, gave birth in the end to materialist thinking.[7] It is erroneous to imagine, however, that empiricist psychology, as the basis either of a complete theory of knowledge or of experimental method in science, had a necessary connection with the type of naturalism here in question. The features essential to the latter were quite unrelated, as we have seen, to Lockean doctrine. How then, it is asked, did Locke come to be so familiarly quoted and so ubiquitously enlisted, by certain philosophes, in the materialist cause—indeed to such an extent, that students of the period have since found it best simply to reecho the philosophes themselves on that score?

First of all, Diderot and his associates, with the abandonment of Cartesian metaphysics, unanimously rejected innate ideas and adopted in their stead a sensationalist theory of knowledge. This, in turn, was to offer them a final crowning argument in favor of human automatism. For, although it

[7] Instances of exaggerating Locke's contribution to French materialism are: W. Windelband, *A History of Philosophy*, trans. J. H. Tufts, New York, 1895, pp. 453-458; John Grier Hibben, *The Philosophy of the Enlightenment*, New York, 1910, pp. 123-126; Oskar Ewald, *Die Französische Aufklärungsphilosophie*, München, 1924, pp. 74-75, 112; B. A. G. Fuller, *A History of Philosophy*, New York, 1945, II, 193-194; Paul Hazard, *La Pensée Européenne au XVIIIe siècle, de Montesquieu à Lessing*, Paris, 1946, I, 159-162; Daniel Mornet, *La Pensée française au XVIIIe siècle*, 7e éd., Paris, 1947, pp. 98-103.

had resulted from the beast-automaton and the *Traité de l'homme* that life, sensibility, sensations, and passions could be regarded as mechanical effects, there still remained a certain indecision about the nature of the purely rational soul—an indecision that might be turned (and often was) to the profit of spiritualist philosophy. However, when Locke refuted innate ideas by showing that even the most abstract and universal notions were ultimate derivatives of sense-experience, the exponents of the man-machine were able to deny that a spiritual principle transcended the mechanistic explanation already held to be valid for all organic functions save cogitation alone. If it is not quite accurate to say that certain of the philosophes were materialists because they were Lockeans, it could more truthfully be said that they were to some extent Lockeans because they were materialists—although they of course had, in common with non-materialists, other reasons as well for embracing sensationalism. Moreover, the career of the *bête machine* itself would suggest that, even without the aid of Locke, Cartesian automatism would, as La Mettrie repeatedly made so plain, have culminated logically in the man-machine. Denesle's judgment, in linking the animal-automaton with the more recent tendencies of his contemporaries, also made this quite explicit: ". . . for if matter, simply by its organization, could be brought to the level of sentiment, then the Materialist will immediately conclude that it can likewise be elevated to that of thought and knowledge."[8]

Still, Locke was made out by some philosophes as a founder of naturalism, and to this end frequent reference was made to his famous conjecture, imprudently popularized by Voltaire's *Lettres philosophiques*, that matter as such, for aught we know and in view of God's infinite power, could conceivably be endowed with the ability to think. So far as materialist discussion in the Enlightenment was involved, this passing

[8] Denesle, *Examen du matérialisme*, I, 389.

observation of the *Essay concerning Human Understanding* became more seriously and fervently an issue than the remainder of the book's contents. A monograph on its picaresque fortunes across the Channel would make both entertaining and instructive reading. But those who have accepted the habitual allusion to this conjecture as proof of a deep influence by Locke on the materialism of La Mettrie, Diderot, Buffon, and D'Holbach have greatly overstated the case. The supposition that, God willing, matter might think, was in itself a proposition entirely indifferent to the principles and structure of Locke's epistemology. This, and the fact that Locke mentioned the idea in the non-affirmative form of an insoluble speculation ought, perhaps, to have aroused caution about pronouncing on the spirit in which the philosophes made capital of the Englishman's remark. As events in France had it, the "possibility of thinking matter" came to be more intimately related to the naturalistic meaning of Descartes's science than to the fundamentals of empiricist psychology; such was, in effect, the secret of its enormous, but otherwise baffling, success in the philosophes' *milieu.* An instance of this, it is remembered, was Diderot's letter to Voltaire dealing with the prophetic insights of Saunderson into the origins of the world and of organic beings.

The philosophes were not, as a rule, too scrupulous about whom they quoted in favor of their opinions, and there were strong tactical reasons for quoting Locke. These were, first, the brilliant reputation of British philosophy in eighteenth-century France and its usefulness as a fulcrum for throwing off balance in science, theology, morals, and politics the *status quo* of the *Ancien Régime.* Second was the happy association of what D'Alembert called Locke's "physique expérimentale de l'âme" with the generally acknowledged superiority of Newton's physics and experimental method. A notable victim of the philosophes' tactics was Condillac who, by his own admission, was passionate about Lockeanism chiefly because he believed in the possibility of rendering an important service to religion by expounding it "in a manner that the Materialists could not make an abuse of."[9] If his efforts were fated to have little adverse effect on the maturing of naturalism among those same persons who, notwithstanding this, were willing enough to accept the Lockean epistemology promulgated by Condillac, the reason was, most likely, that the author of the *Traité des sensations* had erred in assuming that their ideology was based on empiricist premises. There was, to be sure, a type of behavioristic doctrine in the Enlightenment, best elaborated by Helvetius, which did actually grow with logical consistency out of Lockean soil. But the standpoint of Helvetius remained distinct from scientific naturalism despite certain affiliations with it, and even came eventually into open conflict with the specific views advocated by Diderot.

That the materialism of the Enlightenment had affinities with the teaching of Epicurus and Lucretius, as the philosophes themselves and their critics (both old and new) have often maintained, remains, of course, true. Yet the problem is to know in exactly what relationship to ancient Epicureanism the eighteenth-century ideology stood. The results of the present study would indicate that the latter was not, in any strict sense, a direct or faithful offspring of Graeco-Roman tradition, but was instead its outcome only insofar as it had first been reformed and reinvigorated by Descartes, and to a lesser degree by Gassendi. Cartesian science, as embodied in the mechanistic cosmogony of the *Monde* and *Principes*, the automatist theory of organism, and the psychophysiology of the *Traité de l'homme* and *Passions de l'âme*, proved to be the chief medium through which classical thought passed in its transmission to Diderot's epoch. Epicureanism, in filtering through the alembic of Cartesian doctrine, underwent a profound change which, being evidenced also by the

[9] Condillac, *Œuvres philosophiques*, I, 385.

philosophes' naturalistic outlook, effectively sets apart their debt to ancient schools from their debt to Descartes.

The fundamental distinction to be made, in this regard, is that Epicureans had, unlike Descartes and his followers, defined Nature in terms of the concept of Chance, to whose operations (plus a casual assistance from the Gods) the actual state of the universe was attributed. With the notion of Fate at its core, such a materialism was apparently elaborated in the interests, less of physical inquiry, than of a fatalistic metaphysics having a preponderantly moral import. But Descartes's revision of Epicurean ideas in the new context of mechanistic physics extricated the all-important concept of Nature from the sterile rule of contingency, and made it the expression of the fixed and determinable laws of matter in motion. Descartes himself stated the gist of this revolutionary advance. Even if Democritus had been first with the plan to penetrate the mystery of nature by means of the figures, motions, and situations of material particles, Descartes was soon obliged to discard his physics, "because he did not explain in each particular case how all things had been formed by the meeting of these small bodies, or rather, if he explained it of some, the reasons given were not so dependent on one another *that it was made apparent that all of nature could be explained in the same manner*."[10] The Cartesian reform consisted, precisely, in the elimination of the factor of contingency from both the physical and logical bases of science, and in the substitution of law, or necessity, in its place. By this decisive event, the philosophes' scientific naturalism was made possible. D'Alembert perhaps worded best the tribute that the French philosopher had earned, in that regard, from the century of *lumières*: "Let us acknowledge that Descartes, forced to create an entirely new physics, could not have created a better one . . . and that

[10] Descartes, *Principes*, iv, 202; ix, 320.

if he erred about the laws of motion, he was at least the first to have guessed that there must exist such laws."[11]

The new conception of Nature as something independent both of God and of Chance, or rather as an intermediary between the two possessed of an intrinsic finality, found concrete support in the vitalist biology of the 1740's. Calling for the discovery of the "hidden causes in Nature," La Mettrie's *Homme machine* transcended therewith the fruitless chaos-atoms-chance triumvirate of classical Epicureanism.[12] His *Homme plante* affirmed scientific faith in the notion of an "Active Nature" which contrasted, also, with the passive Nature of those who, marvelling at phenomena and lucubrating on natural theology, spent their lives "in counting the little segments of the auditory membrane of certain fish, or in measuring, if you please, the distance that a louse is able to leap."[13] The same definition of Nature, applicable to moving matter in progress from chaos to order, underlay the cosmogonical speculations of Buffon and the transformistic biology of Diderot, both of which were careful to avoid a reliance on finalism and fatalism alike. In a generic sense, the philosophes' scientific naturalism may be termed (as La Mettrie entitled one of his lesser-known writings) a "Système d'Epicure." In its specific character, however, it had a more modern background. There is reason to believe that La Mettrie was being more accurate when he referred, in an ironical context, to the ideology of his age as a "système Epicuro-Cartésien retourné & mal cousu."[14]

Diderot has in these pages been selected as the central figure of his group, for it was in his thought that all the different phases and facets of the period's materialism found their most integrated and synthetic expression. The encyclopedist's ever-curious, mobile mind caught and reflected

11 D'Alembert, "Discours préliminaire," *Œuvres*, I, 67.
12 La Mettrie, *Œuvres philosophiques*, III, 54-55.
13 *ibid.*, II, 22.
14 *Les Animaux plus que machines*, *ibid.*, 33.

all the major themes thus far treated: a hypothesis for the development of the cosmos and organic patterns from the laws of matter in motion (or fermentation); the exclusion of final causes from the province of natural science, and the definitive freeing of the latter from theological controls; the mechanistic view of man; and lastly, transformism. It would not be amiss, now, to summarize the manner and extent of the contributions which Diderot's several colleagues made to the same intellectual trend.

La Mettrie is specially to be credited with having formulated the man-machine thesis as a logical extension of Descartes's animal-automaton and mechanistic physiology. Moreover, the author of *L'Homme machine* used the implications of vitalist data for arriving at the idea of a creative Nature, which promptly became a key-tenet of the materialist sect. It must be put on record, however, that La Mettrie was long unsympathetic towards the theoretical attempt, exemplified by the *Monde*, to derive all things in the world, *tant générales que particulières*, from mechanical causes. Notwithstanding this, in the *Système d'Epicure* of 1750 his thinking would appear, by its new interest in transformistic theory, to have at last been enticed by just such speculations. Thus influenced in his own turn by the advances of a materialism which he had done much to inaugurate, it is difficult to predict what place La Mettrie would have assumed in that school, and if he would not have given his equivalent of the *Rêve de D'Alembert*, had he not died prematurely in 1752 from the after-effects of overindulgence in a pheasant *pâté*.

Buffon deserves recognition mainly for having framed and published, at an early date, the cosmogonical and biological hypotheses in which Diderot and his associates found both incentive for, and corroboration of, their own natural philosophies. But the author of the *Histoire naturelle*, a prudent man in the King's employ, chose not to jeopardize his post as director of the *Jardin du Roi* by agreeing with the materialist and anti-religious conclusions to which, in the hands of others, his scientific achievements quite logically led. A member of D'Holbach's circle, Buffon could hardly have been unaware of the close bond between his own theorizing and the evolutionary materialism of a contemporary such as Diderot. Although he had formally retracted under official pressure the unorthodoxies suggested by the *Théorie de la terre*, Buffon's history of the globe, in altogether dispensing with Christian chronology, became "the refrain of the students and professors of . . . atheism."[15] His sponsoring of spontaneous generation and organic molecules was instrumental in establishing a purely mechanistic and unfinalistic biology in the Enlightenment. Yet, he avoided quite typically the enlargement of such principles into a man-machine viewpoint. Buffon's equivocal role in the rise of naturalism is best seen, perhaps, in the apt remark about him that "he spoke of the human soul to the Sorbonne, and about matter with his friends."[16] Despite such ambiguities, the "Première vue de la nature" stated admirably the concept of a dynamically creative Nature which, while subject to God's will in an ineffable sense, remained "a ceaselessly active worker who knows how to utilize everything; who, working on her own initiative [d'après soi-même] with always the same materials, far from exhausting them, renders them inexhaustible: time, space, and matter are her means, the universe her object, motion and life her ends." Buffon's science illustrated, moreover, that subordination of experimental procedure to a more imaginative method of hypothetical deduction which, inherited from Descartes, was to exert so decisive an influence on Diderot's group. In reference to the "systematist" features of the *Théorie de la terre*, the *Correspondance littéraire* made a valuable admission about the persistence, in

[15] La Harpe, *Cours de littérature*, XVII, 8.

[16] J.-E.-M. Portalis, *De l'usage et de l'abus de l'esprit philosophique durant le dix-huitième siècle*, Paris, 1820, I, 163.

the eighteenth century, of Descartes's methodology: "What is most extraordinary is that the *esprits systématiques* perceive very well the weakness of their comrades for chimera, but are never themselves conscious of being in the same condition."[17] Notwithstanding his caution or timidity on many points, Buffon must be accredited with having furnished both a general outline and a concrete content for scientific naturalism. In this connection, it was the *Histoire naturelle* that drew much of the censure of anti-materialist critics.

D'Holbach, who sought to codify the meaning and spirit of the doctrine held by his colleagues, gave to it—as is perhaps inevitable in codifications—a somewhat rigid, dogmatic, and arid aspect. Overstating its case in the interests of atheistic propaganda, the *Système de la nature* minimized the specifically scientific motivations and *raison d'être* of the naturalism it championed. Instead, D'Holbach busied himself primarily with the moral consequences of a system of mechanistic determinism. Holding that "all the errors of mankind are errors in physics," he voiced the final outcome of the relationship between natural science and ethics that the *Discours de la méthode* had averred in the previous century. The proper business of man became the study of nature: "let him learn its laws, contemplate its energy and the immutable fashion in which it acts; let him use his discoveries for his own felicity. . . ."[18] The great popularity of the *Système de la nature*, furthermore, helped considerably in the diffusion and eventual triumph of the materialist ideas catalogued between its covers.

Maupertuis, strictly speaking, does not belong to Diderot's faction. His leading concern in natural philosophy was to effect, on the example of Leibniz, some sort of compromise between finalistic theology and mechanistic science. This did not prevent Maupertuis from participating indirectly in the

[17] Grimm, Diderot, *et al.*, *Correspondance littéraire*, III, 305.
[18] D'Holbach, *Système de la nature*, I, 2.

growth of scientific naturalism, by virtue of his speculations on vitalist phenomena and transformist process. Maupertuis did not succeed in resolving the conflicts, so clearly perceived and described by him, that sprang from the coexistence in the eighteenth-century of the Newtonian teleological scheme and the naturalistic purport of biology in the 1740's. Instead, his discussions of the subject seemed only to strengthen each of the two viewpoints which continued to remain separate and incompatible. The principal value for us of Maupertuis's thought consists in the bright light it throws on the issues basic to the appearance of the philosophes' materialism.

The case of Montesquieu is a special one. His *Lettres persanes* made known some of the radical implications of the Cartesian physics in which he seemed quite well-versed. It was not unusual, then, that in 1719, while active in the affairs of the *Académie des sciences de Bordeaux*, Montesquieu should have launched a somewhat ambitious "Projet d'une histoire physique de la terre ancienne et moderne," and attempted to secure the collaboration of scientists interested in geology. If with the self-assertion of Montesquieu's genius in another sphere this enterprise came to naught, there was perhaps no serious loss. The principles of Descartes's *Monde* turned up in the general conception of the *Esprit des lois*. It was with the analogy of the laws of nature in mind that the philosophe proposed to explain, as certain contemporaries were engaged in doing for the origination of the cosmos, the evolution of political institutions and of social organisms. While aware that human events did not have the predictability of the physical, Montesquieu's notion of historical law remained nonetheless in close touch with the naturalistic science of his time. In the vision of fixed mechanical causes forming the world independently of both divine Providence and sightless Chance, the *Esprit des lois* had philosophically one of its points of inception:

"Those who have said that *blind Fate has produced all the effects we see in the world*, have uttered a great absurdity; for what greater absurdity than the idea of blind Fate producing intelligent beings? . . . As we observe that the world, formed by the motions of matter and deprived of intelligence, subsists always, it must be that its movements have invariable laws; and if we could imagine a world other than the present one, it would either observe constant rules or be destroyed. Thus the creation, which appears to be an arbitrary act, presupposes rules as invariable as the fatalism of the atheists. It would be absurd to say that the creator, without these laws, could govern the world, since the latter would not subsist without them. Those laws are permanently established relations. Between one body moved and another, it is according to the relations of mass and velocity that all motions are communicated, augmented, diminished, lost; all diversity is *uniformity*, all change *constancy*."[19]

With a definition of law borrowed from the science of nature, Montesquieu explored the causes behind the cyclic rise and decay of civilizations, and published his findings, by a curious coincidence, at almost the same moment that Buffon and Diderot, with the same postulates, were studying the homologous processes of the material order. Under the impact of Cartesianism, the search for origins thus replaced, in political as in physical inquiry, the reliance on traditions. It was because natural science had already become "historic" in outlook that history was enabled, conversely, to become "scientific."

Despite the various substantiations of the Cartesian source for their naturalism, the philosophes remained reluctant, as stated earlier, to avow the extent of their indebtedness. There was a rather simple reason for this. Their ideological heritage had represented, from the start, an unofficial version of Car-

[19] Montesquieu, *Œuvres*, III, 90-91.

tesianism, and, if publicized as such, would have had to compete for authority with the deeply-intrenched official tradition with its apologetic investments in dualism, innate ideas, ontological proof of God's existence, and so forth. Descartes having claimed, moreover, that his physical system was the necessary consequence of certain prior conditions of his metaphysics, a portion of the anti-Enlightenment party had, as La Harpe observed, "put trust in Descartes's physics, because he was a good metaphysician."[20] Under the circumstances, even if his natural philosophy did not depend absolutely on Descartes's metaphysical position, but was entitled logically to evolve in its own right, the philosophes were not the ones to undertake to demonstrate the point. It was easier and wiser to forego the pleasure of being styled Cartesians. It proved more pertinent to the practical aims of *lumières* to adopt a tactic of equivocation, to appropriate from the storehouse of Descartes's thought whatever could be of use, and to relinquish to the opposition the privilege of being technically considered Cartesians. In compensation, the philosophes habitually called themselves, even if often too vaguely, the true heirs of the revolution effected by Descartes—a boast that may be assumed to have applied to a multitude of things.

The eighteenth-century materialists linked their doctrine so frequently with classical Epicureanism that they succeeded, possibly against their better intent, in confusing their critics about its authentic meaning. Taking two typical instances, Bergier's *Examen du matérialisme* and Pluquet's *Examen du fatalisme*, the main objection to the ideology of the epoch is found to consist in the belief, common also to deists like Voltaire and Rousseau, that "blind Chance" could not have engendered the ordered state of the cosmos, nor its regularity of purposeful patterns; that, consequently, it was necessary to explain by the intervention of an Intelligent

20 La Harpe, *Cours de littérature*, IX, 379.

Cause the development and functions of all objects.[21] So far as the defenders of orthodoxy sought thus to catch their antagonists on the horns of a dilemma which they themselves (aided by the philosophes) had manufactured—namely, an either-or alternative between Blind Fate and Intelligent Creator—they were in reality losing sight of the target, and their fastidious critiques were destined, by their irrelevancy to the central issue of the controversy, forever to gather dust. In refusing to come to grips with the new conception of Nature and the related methods and goals of scientific inquiry, such critics persistently jumbled, in a manner profitable to religious control, the separate domains of theology and of physics. Pluquet, for example, was conscious of La Mettrie's definition of Nature as something that was neither God nor Chance, and even quoted *in extenso* the passage from *L'Homme machine* in which this had been offered under the auspices of vitalist speculation. But Pluquet was unable to forget the "Blind Chance-Intelligent Cause" dichotomy: "I have shown that the formation of the polyp, which serves as pretext for the supposition of these occult causes [i.e. La Mettrie's 'hidden causes in Nature'] . . . indicates that this animal could not have been produced by a blind and necessary force. I have therefore had to reject all those physical causes to which some would like to attribute the formation of the world."[22] Against such misrepresentations, D'Holbach countered that scientific naturalism did not conceive of Nature in terms of the Epicurean *Fatum*: "Let no one tell us that according to [our] hypothesis we ascribe everything to a blind cause, to the fortuitous concourse of atoms, to *chance*. We call *blind causes* only those whose behavior, force, and laws are not known to us. . . . This may serve as answer to the eternal objection made against the partisans of Nature, who are accused endlessly of *attributing all things to*

[21] Bergier, *Examen du matérialisme*, I, 65; Pluquet, *Examen du fatalisme, passim.*

[22] *ibid.*, III, 327.

chance."[23] Nonetheless, a profound gulf of incomprehension lay between the possible objectives of natural science and the imagined prerogatives of revealed religion. Seen in this light, the major part of the polemics that raged between the "côterie holbachique" and its Christian or deistic adversaries shows up as a consummate *malentendu*, in which Diderot and his colleagues argued for one thesis, while Voltaire, Rousseau, and others still more orthodox, found themselves allied by expediency in arguing against quite a different one.

The naturalism of La Mettrie, Diderot, Buffon, and D'Holbach was fundamentally a means of specifying and vindicating the rightful authority of science to give, by appropriate methods and principles, the completest interpretation of the physical universe within the compass of intelligibility. As such it was primarily, not a dogmatic affirmation that the "truth" about things had at last been found out and all obscurities dispelled, but rather an initial act of intellectual humility and avowal of ignorance, followed by a proposal to *search* for rationally and empirically verifiable knowledge under conditions that would guarantee the maximum of speculative freedom to the searcher. In eighteenth-century France, this came inevitably into conflict with what authoritarians regarded as dogmas essential to Catholic faith; and the practice then was to label as atheists those who today would perhaps be considered no more than agnostics. But so perceptive a critic as Gautier d'Agoty, while cognizant of the full measure of Diderot's advanced thinking, realized that atheism was not properly a necessary or integral part of naturalistic science: "who is the writer so bold as to reproach M. Diderot that, in saying that Nature was not God, he meant to say that God did not exist?"[24] The conventional type of atheistic materialism was obviously not

[23] D'Holbach, *Système de la nature*, II, 173.

[24] Gautier d'Agoty, *Observations sur l'histoire naturelle*, Xe Partie, p. 60.

quite the same as the ideology of Diderot, "even if, according to him, it is only in Nature that we are able to search for the causes of all motion, and even for the principles of animal life." Resulting from the precept and impetus of Descartes's *Monde* and *Principes*, the science of the Enlightenment, particularly in the cosmogonical and geological hypotheses of Buffon and in the transformistic biology of Diderot, had aimed at giving a mechanical explanation for the emergence of the world and its contents. Such a program, as it happened, could not but have threatened to overthrow the traditional Christian views on the same subject. A theological observer like the abbé Gauchat insisted, with *Telliamed* and the *Théorie de la terre* in mind, that those sciences which investigated the age of the earth and the evolution of things outside the Christian chronology, and framed cosmogonical theories at variance with *Genesis*, were in effect attacking Christianity at its roots.[25] When the abbé Pluquet remarked, in the same spirit, that "the philosopher can neither determine the origins of the world, nor the states through which it has passed, nor the revolutions that are in preparation,"[26] one can be sure that this vow of skepticism was made in the higher interests, not of scientific certitude, but of religious faith.

Yet the philosophes had not stopped even there. As Descartes had projected in the previous century, Diderot and his associates set out to clarify also the process by which organic beings, along with the cosmic environment, had developed from physical causes. To this end they imagined, most strikingly and suggestively, a general hypothesis of transformism. Needless to repeat, ecclesiastical authority could see in such an enterprise little else than an impious assault against the Mosaic "physics"; and accordingly, the heavy hand of dogma bore down upon what was actually a

[25] Gauchat, *Lettres critiques*, xv, 66-67.
[26] Pluquet, *op. cit.*, III, 816-817.

most promising and fertile line of inquiry. Bergier summed up the standpoint of all who deplored the materialist orientation: "The species are . . . invariable and indestructible, and their germs are inalterable: such is the supreme law emanating from the Creator."[27] But with the extension of transformistic ideas to the appearance of man himself, the axiom of spirituality on which the Church founded its conception of human nature seemed also to be menaced: "if men were once fish, if there are still in the sea fish destined to become human beings, one of two things follows: either man does not have a spiritual and immortal soul, or else fishes have the same soul—two equally impious suppositions."[28] The evolutionist thesis logically combined, moreover, with the metamorphosis of Descartes's *bête machine* into the man-machine, curtailing further the uses of the spiritual and immortal soul in the task of studying and defining man. The Church saw in this complex of scientific naturalism a formidable offensive mounted against itself, and responded by denouncing as atheists all and sundry who sympathized with the new cosmogonical, biological, and physiological trends. Its failure to distinguish between what was rightly a part of religious belief, and what lay outside its competence to judge, produced as counterpart the inclusion of strongly-phrased, but unessential, anti-religious or atheistic elements in the doctrine of certain philosophes. Under the pressure exerted by the Church, which in the *Ancien Régime* could command against "subversives" the punitive power of the State, it was hardly possible to hope for a sober examination and honest settlement of the differences existing between the two opposing sides. Diderot knew that "the method of theologians is first to anathematize new opinions, and afterwards

[27] Bergier, *Examen du matérialisme*, I, 148.

[28] Gauchat, *op. cit.*, xv, 224. The deistic equivalent of this general *critique* may be read in the "Profession de foi du vicaire savoyard," which Rousseau inserted in *Emile*.

to reconcile them with their dogmas."[29] It was normal that materialist science, like the general movement of Enlightenment of which it was a major aspect, should have found itself engaged in a dramatic struggle with the established faith. Out of this unfortunate background, the modern antithesis between science and religion has in large measure taken shape.

Correctly understood, scientific naturalism was not concerned with a denial of the existence of God, the agency of an ineffable Providence, and similar concepts peculiar to metaphysics, ethics, or theology. It sought basically to prevent the use, or more exactly, the abuse of such notions in the investigation of nature. In doing so, it guaranteed to scientific inquiry the maximum degree of autonomy in the explanation of physical phenomena. This was, perhaps, the principal contribution of La Mettrie, Diderot, and Buffon to the history of science. The school of eighteenth-century naturalism, having its sources in Cartesian tradition, succeeded in formulating the ideal of an entirely independent and mature philosophy of nature, and in creating the ideological atmosphere in which rational inquiry has since been free to give the best account of man and the world of which it has been capable. There is reason today, as compared with the nineteenth century, to limit our faith in the complacent pretensions to progress made in the name of science; but in the eighteenth century the cause of progress was best to be served by giving the widest possible scope to its possibilities. It was then necessary to assume that Nature had come before religion, mythology, and art. But the broad range fixed for the activity of a science of nature by Diderot and his colleagues has by no means been exhausted or by-passed in the course of time. The effort to solve the cosmogonical, biological, and psycho-physical problems out of which, in the inadequate context of thermodynamics, the philosophes' nat-

[29] Diderot, xvi, 123.

uralism had taken form, still defines, in an ultimate and compelling sense, the vocation of Diderot's intransigeant interpreter of nature.

But Diderot's thought is still valid for us also by its having expressed, not only the widest theoretical scope of natural science, but its limitations as well. In the study of man, which Diderot's humanism took to be the proper end of all inquiry and knowledge, the role of physics was supplemented and often replaced, unlike the narrower practice of the *Système de la nature*, by considerations and methods that did not have immediate reference to the laws of matter in motion. Forestalling the claims of a shallow scientism, Diderot was aware that the man-machine physiology, transformistic biology, and related doctrines did not in fact completely define man, nor convincingly interpret his peculiar destiny. *Jacques le fataliste* was a satire on the kind of fatalism which, fusing scientific pretentiousness with a disenchanted reading of Spinoza's chapter on "Human Bondage," might seek entirely to identify human values and purposes with the inexorable chain of physical event. The preamble to the *Interprétation de la nature* had cautioned the reader against believing either that man was a machine, or that Nature was God.[30] Even La Mettrie conceded that the man-machine hypothesis, whatever its usefulness in analysis, did not exhaust its object;[31] concerning man's ultimate destiny, the *Homme machine* acknowledged that the problem lay outside the jurisdiction of natural science.[32] In a still more positive manner, the *Neveu de Rameau* and the mass of Diderot's esthetic criticism appealed to a set of values, both in art and in morals, which transcended—even if willing to utilize so far as possible—the principles on which the philosophe's system of nature was erected. Historians of ideas have frequently

30 *ibid.*, II, 7.
31 La Mettrie, *Œuvres philosophiques*, III, 6-7.
32 *ibid.*, 88-89.

seen in Diderot's esthetic and moral writings a deep contradiction of his materialism and have accordingly accused him of glaring inconsistency. Such criticism has failed, however, to understand that his naturalism did not propose a total interpretation of reality on the level of a metaphysics, but sought essentially to establish the autonomy of natural science in the investigation of the physical world. Once this is realized, there is no inconsistency. Beyond the reaches of a science of nature, Diderot was concerned with discovering and describing—as Descartes, too, had done—a moral liberty subjective to man in the midst of an objective determinism. Diderot's thought, in this respect, still has meaning for us. Having gotten over the verbal tyrannies promulgated by nineteenth-century squabbling between idealists and positivists, it is now possible to maintain that moral freedom and physical necessity are not antithetical concepts, but instead are complementary; that choice of action and creativity of act are possible only under given conditions of fixed and immutable law; that the determinism of the natural order is, in actuality, the solid bedrock on which we are able to construct—and such is the true sense of moral freedom—what is an expression specifically of our own values and ends; and that moral liberty, therefore, need no more mean arbitrariness of behavior than physical necessity need imply human bondage. Diderot's thinking, taken in its integrity, is a prolegomenon to such a view of man's situation in the world; viewed in a latter-day light, the equally intense activity of his intellect in the different spheres of science, ethics, and art represents the achievement of what Sainte-Beuve justly described as "le génie le plus synthétique de son siècle."[33]

The naturalism of Diderot remains, by its conception of the role and prerogatives of science as well as by its approach to the moral scrutiny of man, consistent with what might

[33] Sainte-Beuve, in an article appearing in the *Globe*, 20 Sept. and 5 Oct. 1830.

be called the tradition of liberalism. It is an excellent illustration of the liberal answer to two questions inextricable from the entire shaping of Western civilization: the problems of free inquiry and of the individual.

The ideology championed by La Mettrie, Diderot, Buffon, and D'Holbach, as we have seen, was motivated by the wish to liberate the sciences of nature from the tutelage of dogmas having intrinsically nothing to do with the methodical, systematic investigation of the facts and processes of the physical universe. The main obstacle to the maturing of natural science happened, during the age of *lumières*, to be (for historical reasons rooted in the culture of the Middle Ages) the Church with its two-pronged weapon of supernatural theology and all-too-secular politics. Scientific naturalism thus owed its meaning and force to the justifiable reaction against what the philosophes' critics persisted in calling "la saine Philosophie," or what might be designated as "science in the service of piety." The substitution for this of other forms of "religious" right-thinking in the twentieth century, with the demand that astronomy, physics, biology, genetics, etc., confirm and corroborate new dogmas about man's nature and destiny, would be in effect a reversal of the ideological advance from which the materialism of Diderot's circle ensued. Such a subordination of free inquiry to considerations alien to the methods and postulates of natural science would represent a manifest regress to the state of affairs against which modern thought, from the Italian Renaissance, through Descartes, to the philosophes, had to contend. The message central to the present phase of eighteenth-century *lumières* deserves still to be pondered, if the hard-won privileges of science, happily rescued from the protracted domination of theology, are not to fall into other equally damaging toils, and if the unbiased exploration of the physical universe is to make its maximum contribution to man's desire, as Descartes said, of "seeing his way clearly in this world."

Diderot's naturalism implied, moreover, a conception of man based on the recognition of the ultimate reality and dignity of the individual. This is best made apparent by the differentiae which Diderot himself set up between his own doctrine and its closest contemporary rival. As Karl Marx has stated,[34] it was the current of epistemological ideas issuing from Hobbes and Locke which, acclimatized in France by Voltaire and Condillac, had there produced the behavioristic materialism of Helvetius; whereas the progress of Cartesian tradition, developing separately, had already brought forth in the same *milieu* a materialism reposing fundamentally on a philosophy of nature. These two intellectual complexes contained, actually, two quite divergent approaches to the problem of individuality. The point was not fully realized, however, until the distinct streams of Lockean and Cartesian materialism, which up to a late date often pooled resources in the common struggle (Diderot himself is said to have composed some pages of Helvetius's works), came at last into open and conscious disagreement. The detailed expression of this conflict was Diderot's lengthy, pungent *Réfutation de l'ouvrage d'Helvétius intitulé L'Homme*, and a similar short criticism of the same author's *De l'Esprit*.

A single persistent argument runs through the whole of the *Réfutation d'Helvétius*, composed around 1773 to 1774. This is Diderot's insistence on the essential role, in psychological and moral theory, of the physical organization, as opposed to Helvetius's almost unilateral reference to sensualist criteria. Whereas Diderot's orientation in a "science of man" is thus seen to derive from the history of physiological thought that had converted Descartes's beast-machine into the man-machine, the doctrine of Helvetius was an outgrowth, as Diderot clearly realized, of the quite divergent

[34] Marx, *Die Heilige Familie*, vi, iii; Werke, Abt. i, Bd. 3, p. 307.

effort of Lockean empiricism to study the human being in terms of the external objects and events acting upon him through sense-experience.[85] Contesting neither the positive truths of Locke's epistemology, nor the many astute observations suggested by it in *De l'Homme*, Diderot nonetheless rejects the general proposition that man is best definable by such principles: "one of the greatest inconsistencies of [Helvetius] is that he makes the difference between man and beast consist in the diversity of their organizations, and excludes this cause when it is a question of explaining the difference from one man to another."[86] For Diderot, the psychophysical parallelism associated with Descartes's mechanistic physiology was the most dependable starting-point for the determination of human traits and types.[87] The metaphysical apriorism of Descartes's innatist psychology had been assimilated to the physiological apriorism of the *Traité de l'homme*, that is, to the priority of the corporeal mechanism over experience proper. Diderot concluded: "A good or bad organization constitutes a difference among men that perhaps nothing could make up for. The anatomists, physicians, and physiologists will demonstrate it to you by an infinite number of phenomena. Open their works, and you will see that the spring, whatever it might be, controlling all our intellectual operations is affected in an almost miraculous manner by the slightest alteration occurring in the remainder of the machine."[88]

And so, not only is one led by way of the *homme machine* to the principles underlying Descartes's notion of organism, but the man-machine has become for Diderot the basis in nature for asserting the primary uniqueness of each individual. Against the "totalitarian" supposition, central to Helvetius's treatise, that "tout homme est également propre à

[85] Diderot, II, 295ff. [86] *ibid.*, 344.

[87] ". . . quelle correspondance plus rigoureuse que celle de l'état de mon corps avec l'état de mon esprit?" *ibid.*, 405.

[88] *ibid.*, 365.

tout," Diderot will insist vigorously that no moral or social system can with justice overlook, in favor of experiential or environmental factors, the concrete primacy of diversities in organization, and hence in personality, among human beings. Diderot even held out the hope, contrary to the Lockean premises prevalent in his day, of deducing a complete psychology from the data of the physical organization, thus indicating the logical conclusion of the line of inquiry instituted by the *Passions de l'âme.* Alluding to Helvetius, he remarked: "If, starting out from the single phenomenon of 'physical sensibility,' either as a general property of matter or as the result of organization, he had deduced from it with clarity all the operations of the understanding, he would have done something original, difficult and beautiful."[39]

In connection with this, Diderot vindicates throughout the *Réfutation* the belief in personal self-determination against the abuse made by Helvetius of the concept of *hasard* in treating of human prospects. The empirical materialism (if it may be so called) of *De l'Homme* fell quite irresistibly into the kind of fatalist attitude that made of the individual an altogether too passive object of external forces and events. For Diderot the concept of Chance, against the background of natural science, had by comparison a minor, even negligible, moral significance. Given the fortuitous circumstance, for example, that had led Diderot to suggest to Rousseau the novel viewpoint of the latter's famous *First Discourse,* in no sense did it follow that a reversal of roles would have made a Rousseau out of Diderot. "Rousseau did what he had to do, because he was he. I would have done nothing or else something quite different, because I would have been myself."[40] If in the *Réfutation* Diderot's determinism opposed itself radically to a doctrine of moral fatalism, this was because Descartes's mechanistic physiology had become the basis for the self-determination of the individual personality. With

39 *ibid.,* 301.

40 *ibid.,* 285.

the dispelling of metaphysical dualism, moreover, it had become possible for physical necessity and moral liberty to fuse and harmonize in the expression of the individual will; the problem of moral freedom, as posed in the context of the *Passions de l'âme*, was resolved as part of the general transformation of Cartesian science into eighteenth-century naturalism. Physical process, individual liberty, and universal change merged in Diderot's thinking: "Everything that takes place in us does so because we are ourselves, always ourselves, and never for a moment the same."[41]

The profound dissimilarity between the two conceptions of man—one the Cartesian-Diderot, the other the Lockean-Helvetius—must be ascribed, ultimately, to that difference of first principles between Descartes and Locke to which the philosophes were heirs, and which they tried in vain to bridge. It ensued from the position of Helvetius that man was definable in terms of what lay outside him: he was the product of his total physical and social environment, education, and so forth. For Diderot, man was essentially understandable in relation to his physical organization. This latter was *a priori* in that it regularly preceded and continually modified experience, and "innate" in the sense that, given the identity of matter and soul in the man-machine theorem, the organization became the man himself. The behaviorism of Helvetius represented, consequently, an extrinsic determination of the human being: the determinist physiology of Diderot represented an intrinsic necessity. The thesis of *De l'Homme* rested, in the last analysis, on the assumption of a passive *tabula rasa*: the man-machine was founded on the conception of an Active Nature, of which the individual was a particular expression and, like Nature itself, self-determining by inherent laws. The man of Helvetius retained the moral anonym-

[41] *ibid.*, 373. D'Holbach's *Système de la nature* (I, 195) placed the question of personal freedom on the same basis: "Exiger d'un homme qu'il pense comme nous, c'est exiger qu'il soit organisé comme nous. . . ."

ity and passivity of the *tabula rasa*: the man of Diderot possessed the creative dynamism and internal finality of Nature itself.

The juxtaposition of Diderot's views with those of his contemporary makes plain, in reference to the loose composite of eighteenth-century materialisms, to what extent the former was ready to recognize in man the free agent responsible for the devising of his own values and peculiar cosmic destiny. It has long been claimed that Diderot's natural philosophy, picturing humanity merely as a transitory fragment of a necessary and impersonal universe of fermenting matter, was in conflict with the aesthetic and moral sensibilities of his irrepressibly artistic and humanistic genius. An opinion so erroneous has been owing to an imperfect estimate of his scientific naturalism—its motives, significance, and limits. Diderot's materialist ideology, as examined in the foregoing pages, appears to have been one of the earliest and most courageous attempts in the modern era to demonstrate, without sacrificing one to the other, the fundamental sequence and harmony between what lies outside of man and what lies within him.

BIBLIOGRAPHY

Alembert, Jean Le Rond d' *Œuvres complètes,* Paris, 1821-1822. 5 vols.
"De l'Abus de la critique en matière de religion"
"Discours préliminaire de l'Encyclopédie"

Argens, Jean Baptiste de Boyer, marquis d' *La Philosophie du Bon Sens,* nouv. éd., La Haye, 1768. 3 vols.
Lettres juives, ou correspondance philosophique, historique & critique, nouv. éd., La Haye, 1738. 6 vols.

Baillet, Adrien *La Vie de Monsieur Des-Cartes,* Paris, 1691. 2 vols.

Bayle, Pierre *Dictionnaire historique et critique,* Paris, 1820. 16 vols. Articles: Démocrite, Leucippe, Aristote, Ovide, Rorarius, Dicéarque.

Bergier, Nicolas-Sylvestre *Examen du matérialisme: ou Réfutation du Système de la nature,* Paris, 1771. 2 vols.

Bernier, François *Abrégé de la philosophie de Gassendi,* Lyon, 1684. 6 vols.

Beyer, Charles J. "Du Cartésianisme à la philosophie des lumières," *Romanic Review,* 34 (Feb. 1943), pp. 18-39.

Bouillier, Francisque *Histoire de la philosophie cartésienne,* 3e éd., Paris, 1868. 2 vols.

Boullier, David-Renaud *Discours philosophiques sur les causes finales; sur l'inertie de la matière; sur la liberté des actions humaines,* Amsterdam, 1759.

Bouquet, H. L. *L'Ancien Collège d'Harcourt et le Lycée Saint-Louis,* Paris, 1891.

Boureau-Deslandes, André-François *Pigmalion, ou la statue animée,* Londres, 1741.

Bréhier, Emile *Histoire de la philosophie,* Paris, 1930; Tome II, fascicule 2, "Le Dix-huitième siècle."

Brucker, Johann Jakob *Historia critica philosophiae a mundi incunabilis ad nostram usque aetatem deducta,* Lipsiae, 1742-1744; T. IV, part ii.

Brunet, Pierre *L'Introduction des théories de Newton en France au XVIIIe siècle: Avant 1738,* Paris, 1931.

Brunetière, Fernand "Jansénistes et Cartésiens," *Etudes critiques sur l'histoire de la littérature française,* 4e série, 1889, pp. 111-178.

Bruno, Antonino *Cartesio e l'Illuminismo*, Bari, 1949.

Buffon, Georges Louis Leclerc, Comte de *Œuvres complètes*, éd. Flourens, Garnier, Paris, 1853-1855. 12 vols.
"De la manière d'étudier et de traiter l'Histoire Naturelle"
"Discours sur le style"
Histoire des animaux
Théorie de la terre

Burnet, Thomas *Sacra Telluris Theoria*, London, 1681.

Busco, Pierre *Les Cosmogonies modernes*, Paris, 1924.

Cantecor, G. "A quelle date Descartes a-t-il écrit 'La Recherche de la Vérité,'" *Revue d'Histoire de la Philosophie*, II (1928), pp. 254-289.

Cassirer, Ernst *Descartes*, Stockholm, 1939.
Die Philosophie der Aufklärung, Tübingen, 1932.

Castel, Louis-Bertrand (Père) *Le Vrai système de Physique generale de M. Isaac Newton, exposé et analysé en parallèle avec celui de Descartes*, Paris, 1743.

Clarke, Dr. Samuel *A Collection of Papers, which passed between the late Learned Mr. Leibnitz, and Dr. Clarke*, London, 1717.

Condillac, Etienne Bonnot de *Œuvres philosophiques*, éd. G. Le Roy, Paris, 1947. 2 vols.
Essai sur l'origine des connaissances humaines
Traité des animaux
Traité des sensations
Traité des systèmes

Cousin, Victor *Fragments de Philosophie cartésienne*, Paris, 1852.

Coyer, abbé Gabriel-François "Lettre au R. P. Berthier sur le matérialisme," *Œuvres complettes*, Paris, 1782. 7 vols.

Cru, R. Loyalty *Diderot and English Thought*, New York, 1913.

Cudworth, Ralph *The True Intellectual System of the Universe: wherein all the reason and philosophy of atheism is confuted*, ed. T. Birch, New York, 1838. 2 vols. (1st ed. 1678)

Cyrano de Bergerac, Savinien de *Les Œuvres libertines*, précédées d'une notice biographique par Frédéric Lachèvre, Paris, 1921. 2 vols.

Dal Sasso, A. "La Influenza di Cartesio sulla formazione dell'-Illuminismo francese," *Cartesio: terzo centenario del "Dis-*

corso del Metodo." Pubblicazione dell' Università cattolica del Sacro Cuore, Milan, 1937, pp. 227-238.
Dambésieux, abbé *Reflexions sur la physique moderne; ou, la philosophie newtonienne comparée avec celle de Descartes,* Paris, 1757.
Daniel, Gabriel (Père) *Voyage du Monde de Descartes,* nouv. éd., Paris, 1703.
Denesle *Examen du matérialisme,* Paris, 1754. 2 vols.
Derham, William *Physico-Theology,* 12th ed., London, 1754 (1st ed. 1713).
Descartes, René *Œuvres,* éd. Adam et Tannery, Paris, 1897-1913. 12 vols.
(*Traité de la formation du foetus*). *La Description du corps humain*
Méditations métaphysiques
Le Monde ou Traité de la lumière
Primae cogitationes circa generationem animalium
Principes de la philosophie
"Recherche de la vérité"
Traité de l'homme
Traité des passions de l'âme
Discours de la méthode, texte et commentaire par E. Gilson, Paris, 1925.
Regulae ad directionem ingenii, éd. et trad. G. Le Roy, Paris.
Desmaizeaux, Pierre *Recueil de diverses pièces, par Mrs. Leibniz, Clarke, Newton,* Amsterdam, 1720. 2 vols.
Diderot, Denis *Œuvres complètes,* éd. Assézat et Tourneux, Paris, 1875-1877. 20 vols.
Art. "Chaos"
Eléments de physiologie
Entretien d'un philosophe avec la maréchale de ***
Art. "Epicuréisme"
Les Bijoux indiscrets
Mémoires sur différents sujets de mathématiques
Art. "Mosaïque et Chrétienne, Philosophie"
Neveu de Rameau
Pensées sur l'interprétation de la nature
"Principes philosophiques sur la matière et le mouvement"
Promenade du sceptique
Réfutation de l'ouvrage d'Helvétius intitulé L'Homme

Lettre sur les aveugles, éd. crit. Robert Niklaus, Droz, Genève, 1951.
Pensées philosophiques, éd. crit. Robert Niklaus, Droz, Genève, 1950.
Dieckmann, Herbert "The Influence of Francis Bacon on Diderot's '*Interprétation de la nature*,'" *Romanic Review*, XXXIV, no. 4 (1943), 303-330.
anon. *Dissertation sur la formation du monde*. Par l'auteur du Traité des Erreurs populaires, 1738. MS Mazarine 1168.
Dufour, abbé *L'Ame ou le sisteme des Matérialistes, soumis aux seules lumières de la raison*, Avignon, 1759.
(Diderot, D'Alembert, *et al.*) *Encyclopédie, ou Dictionnaire raisonné des sciences, des arts et des métiers*, Paris, 1751-1765. Articles: Cartésianisme (abbé Pestré), Hypothèse (abbé de la Chappelle), Système (D'Alembert).
Estève, Pierre *L'Origine de l'univers expliquée par un principe de la matière*, Berlin, 1748.
Fabre de Charrin *Eloge de Descartes*, Paris, 1765.
Fontenelle, Bernard Le Bovier de *Œuvres*, nouv. éd., Paris, 1766. 11 vols.
Dialogues sur la pluralité des mondes
"Eloge de Malebranche"
"Eloge de Newton"
Théorie des tourbillons cartésiens, avec des réflexions sur l'attraction
Frankel, Charles *The Faith of Reason*, New York, 1948.
Gauchat, abbé Gabriel *Lettres critiques, ou analyse et réfutation de divers écrits modernes contre la Religion*, Paris, 1755-1763. T. XV. (1761).
Gautier d'Agoty, Jacques *Observations sur l'histoire naturelle, sur la physique et sur la peinture*, Xe partie, Paris, 1754. XIe partie, 1754.
Gewirtz, Alan "Experience and the non-Mathematical in the Cartesian Method," *Journal of the History of Ideas*, II (1941), pp. 183-210.
Gouhier, Henri *Essais sur Descartes*, Paris, 1937.
La Pensée religieuse de Descartes, Paris, 1924.
"Sur la date de la Recherche de la Vérité de Descartes," *Revue d'Histoire de la Philosophie*, III (1929), pp. 296-320.
Grimm, Diderot, Raynal, Meister, *et al.* *Correspondance lit-*

téraire, philosophique et critique, éd. M. Tourneux, Paris, 1877-1882. 16 vols.

Guidi, abbé Louis *Entretiens philosophiques sur la religion,* Paris, 1771.

Guyénot, Emile *Les Sciences de la vie aux XVIIe et XVIIIe siècles. L'idée d'évolution,* Paris, 1941.

Histoire de l'Académie royale des Sciences, 1708. 4°

Hoffmann, A. "Die Lehre von der Bildung des Universums bei Descartes in ihrer geschichtlichen Bedeutung," *Archiv für Geschichte der Philosophie,* Bd. 17 (1903), pp. 237-271, 371-412.

Holbach, Paul Heinrich Dietrich, Baron d' *Systême de la nature, ou des Loix du monde Physique & du monde moral,* Londres, 1771. (1st ed. 1770). 2 vols.

Joseph, H. W. B. *Lectures on the Philosophy of Leibniz,* Oxford, 1949.

Jourdain, Charles *Histoire de l'Université de Paris, au XVIIe et XVIIIe siècle,* Paris, 1888. 2 vols.

Kant, Immanuel (*Kosmogonie*) *Allgemeine Naturgeschichte und Theorie des Himmels, oder Versuch von der Verfassung und dem mechanischen Ursprung des ganzen Weltgebäudes nach Newtonischen Grundsätzen abgehandelt.* Herausgegeben von A. J. von Gettingen, Leipzig, 1898. (1st ed. 1755)

Keeling, S. V. "Cartesian Mechanism," *Philosophy,* vol. 9 (Jan. 1934), pp. 51-66.

Keranflech, Charles-Hercule de *L'Hypothèse des petits tourbillons, justifiée par ses usages,* Rennes, 1761.

Observations sur le cartésianisme moderne, 1774.

Krakeur, L. G. and Krueger, R. L. "The Mathematical Writings of Diderot," *Isis,* XXXIII (1941), pp. 219-232.

La Harpe, Jean François *Philosophie du Dix-huitième siècle*; in *Cours de Littérature ancienne et moderne,* Paris, 1825-1826, vols. XVII & XVIII.

La Mettrie, Julien Offray de *Œuvres philosophiques,* Amsterdam, 1774. 3 vols.

Histoire naturelle de l'âme

Les animaux plus que machines

L'Homme machine

Système d'Epicure

Lamoignon-Malesherbes, Chrétien-Guillaume *Observations de*

Lamoignon-Malesherbes sur l'Histoire naturelle de Buffon et Daubenton, Paris, An VI (1798). 2 vols.
Lamy, Dr. Guillaume *De Principiis rerum,* Paris, 1669.
Explication méchanique et physique des fonctions de l'âme sensitive, ou des sens, des passions et du mouvement volontaire, Paris, 1678.
Lange, Friedrich Albert *History of Materialism,* trans. E. C. Thomas, 2nd ed., London, 1879-1881. 3 vols.
Lanson, Gustave "Origines et premières manifestations de l'esprit philosophique dans la littérature française de 1675 à 1748," *Revue des cours et conférences,* 1907-1909.
Laporte, Jean *Le Rationalisme de Descartes,* Paris, 1945.
Lefebvre, Henri *Diderot,* Paris, 1949.
Leibniz, Gottfried Wilhelm *Die philosophischen Schriften,* herausgegeben von C. J. Gerhardt, Berlin, 1875-1899.
Lelarge de Lignac, abbé Joseph-Adrien *Lettres à un Amériquain sur l' "Histoire naturelle, générale et particulière" de M. de Buffon,* Hambourg, 1751. 5 vols.
Lenoble, Robert *Mersenne, ou la naissance du mécanisme,* Paris, 1943.
Locke, John *An Essay concerning human understanding,* ed. A. C. Fraser, Oxford, 1894. 2 vols.
Louis, Antoine *Essay sur la nature de l'ame, où l'on tâche d'expliquer son union avec le Corps, & les loix de cette union,* Paris, 1747.
Lovejoy, Arthur O. *The Great Chain of Being,* Cambridge (Mass.), 1936.
Luc, J. *Diderot,* Paris, 1938.
Luppol, I. K. *Diderot: ses idées philosophiques,* trad. V. et Y. Feldman, Paris, 1936.
Maillet, Benoît de *Telliamed, ou Entretiens d'un Philosophe Indien avec un missionaire françois sur la Diminution de la Mer, la Formation de la Terre, l'Origine de l'Homme,* Amsterdam, 1748. 2 vols.
Maritain, Jacques *Le songe de Descartes,* Paris, 1932.
Marx, Karl *Die Heilige Familie; Werke,* Berlin, 1927-1935.
Maupertuis, Pierre Louis Moreau de *Œuvres,* Lyon, 1756. 4 vols.
Dissertatio inauguralis metaphysica de universali naturae systemate (Dr. Baumann); later published as *Système de la nature, ou Essai sur la formation des corps organisés.*
Essai de cosmologie

Vénus physique
Meslier, Jean *Le Testament de Jean Meslier*, éd. Rudolf Charles, Amsterdam, 1864. 3 vols.
Milhaud, Gaston *Descartes savant*, Paris, 1921.
Mirabaud, Jean-Baptiste de *Le Monde, son origine et son antiquité*, Londres, 1751.
Monchamp, Georges *Histoire du Cartésianisme en Belgique*, Bruxelles, 1886.
Montesquieu, Charles Louis de Secondat, Baron de *Œuvres complètes*, éd. Laboulaye, Paris, 1875-1879. 7 vols.
De l'Esprit des lois
Lettres persanes
"Observations sur l'histoire naturelle"
Montfaucon de Villars, abbé Nicolas-Pierre-Henri *La Suite du comte de Gabalis, ou Nouveaux entretiens sur les sciences secretes, touchant la nouvelle philosophie*, Amsterdam [1715]. (1st ed. 1708)
More, Henry *Enchiridion Metaphysicum: sive, de rebus incorporeis succincta & luculenta dissertatio*, London, 1671.
Moreau, J. N. *Nouveau mémoire pour servir à l'histoire des cacouacs*, Amsterdam, 1757.
Mornet, Daniel *Diderot, l'homme et l'œuvre*, Paris, 1941.
Murdoch, Ruth T. *Newton's Law of Attraction and the French Enlightenment*, 1950. Microfilm copy in Columbia University Library.
Mursell, J. L. "The Function of Intuition in Descartes' Philosophy of Science," *Philosophical Review*, 1919, pp. 391-409.
Naville, P. *Paul Thiry d'Holbach et la philosophie scientifique au XVIIIe siècle*, Paris, 1943.
Needham, John Turberville *Nouvelles Observations microscopiques*, Paris, 1750.
Newton, Isaac *The Mathematical Principles of Natural Philosophy*, trans. into English by Andrew Motte, London, 1729. (1st ed. 1687)
Nollet, abbé Jean-Antoine *Leçons de physique expérimentale*, Paris, 1743-1748. T. II.
Nonnotte, abbé Claude-François *Dictionnaire philosophique de la religion*, Besançon, 1774. 4 vols.
anon. *Nouvelles libertés de penser*, Amsterdam, 1743.
Palmer, R. R. *Catholics and Unbelievers in 18th-century France*, Princeton, 1939.

Pascal, Blaise *Pensées et Opuscules,* éd. Brunschvicg, Paris (n.d.).

Paulian, Aimé-Henri *Traité de Paix entre Descartes et Newton,* Avignon, 1763. 3 vols.

Perrault, Claude *Essais de physique, ou Recueil de plusieurs traités touchant les choses naturelles,* Paris, 1680-1688. 4 vols.

Pluche, abbé Noël-Antoine *Histoire du Ciel, considéré selon les idées des poëtes, des philosophes, et de Moïse,* Paris, 1739. 2 vols.

Spectacle de la nature, Paris, 1732-1750. 8 vols.

Pluquet, abbé François-André-Adrien *Examen du fatalisme,* Paris, 1757. 3 vols.

Polignac, Cardinal de *Anti-Lucrèce, poëme sur la religion naturelle,* trad. par M. de Bougainville, Paris, 1750. 2 vols.

Pommier, Jean *Diderot avant Vincennes,* Paris, 1939.

Portalis, J.-E.-M. *De l'usage et de l'abus de l'esprit philosophique durant le dix-huitième siècle,* Paris, 1820. 2 vols.

Rapin, René *Reflexions sur la Physique; Œuvres,* Amsterdam, 1709. 2 vols.

Ray, John *The Wisdom of God manifested in the Works of the Creation,* 2nd ed., London, 1692. (1st ed. 1691)

anon. *Recherches curieuses de philosophie, ou dissertation sur les principes des choses naturelles, dans laquelle, par le secours d'une méthode nouvelle, on traite de la génération des hommes, des animaux, des arbres, des plantes, de la formation du monde et de sa durée, des causes des vents, du tonnerre, de la foudre, de l'esprit, du raisonnement,* par T.S.J.F., 1713. MS Bibliothèque nationale, fonds français 9107.

Régis, Pierre Sylvain *Système de philosophie, contenant la Logique, la Métaphysique, la Physique et la Morale,* Paris, 1690. 3 vols.

Regius, Henricus *Fundamenta physices,* Amstelodami, 1646.

Regnault, Noël (Père) *L'Origine ancienne de la physique nouvelle,* Paris, 1734. 3 vols.

Rohault, Jacques *Entretiens sur la philosophie,* Paris, 1671.

Rollin, Charles *Œuvres complètes,* éd. Letronne, Paris, 1821-1825. T. XXVIII.

Rosenfield, Leonora C. *From Beast-Machine to Man-Machine,* New York, 1941.

Rousseau, Jean-Jacques *Emile, ou de l'Education,* nouv. ed., Paris, Garnier, 1924.

Russell, Bertrand *History of Western Philosophy,* London, 1946.

Saint-Hyacinthe, Thémiseul de *Recherches philosophiques, sur la nécessité de s'assurer par soi-meme de la verité; sur la certitude de nos connoissances; et sur la nature des etres,* Londres, 1743.

Sennemaud, abbé Pierre *Pensées philosophiques d'un citoyen de Montmartre,* La Haye, 1756.

Sortais, Gaston (Père) "Le Cartésianisme chez les Jésuites français au XVIIe et XVIIIe siècle," *Archives de Philosophie,* Vol. VI, cahier iii (1929).

Souilhé, Joseph *La Philosophie chrétienne de Descartes à nos jours,* 1934.

Stock, Hyman *The Method of Descartes in the Natural Sciences,* New York, 1931.

Taine, Hippolyte *Les Origines de la France contemporaine: l'Ancien Régime,* 16e éd., Paris, 1891.

Terrasson, Jean *La Philosophie applicable à tous les objets de l'esprit et de la raison,* Paris, 1754.

Thielemann, Leland James *The Tradition of Hobbes in Eighteenth-century France,* 1950. MS in Columbia University Library.

Thomas, Antoine-Léonard *Eloge de Descartes,* Paris, 1765.

Tiphaigne de la Roche, Charles-François *Bigarrures philosophiques,* Amsterdam & Leipsik, 1759. 2 vols.

Toland, John *Letters to Serena,* London, 1704.

Torlais, Jean *Réaumur: un esprit encyclopédique en dehors de l'Encyclopédie,* Paris, 1936.

Torrey, Norman L. "Voltaire's Reaction to Diderot," *Publications of the Modern Language Association of America,* L (1935), pp. 1107-1143.

anon. *Traité des trois imposteurs,* Yverdon, 1768.

Turgot, Anne-Robert-Jacques *Histoire des progrès de l'esprit humain; Œuvres,* Paris, 1808. 9 vols.

Vartanian, Aram "From Deist to Atheist: Diderot's Philosophical Orientation, 1746-1749," *Diderot Studies,* ed. O. E. Fellows & N. L. Torrey, Syracuse University Press, 1949, pp. 46-63.

"Trembley's Polyp, La Mettrie, and 18th-century French

Materialism," *Journal of the History of Ideas*, vol. xi, no. 3 (June 1950), pp. 259-286.

Venturi, Franco *La Jeunesse de Diderot*, Paris, 1939.

Voltaire *Œuvres complètes*, éd. Moland, Paris, 1877-1883. 52 vols.

"Dialogues d'Evhémère"

Dissertation sur les changements arrivés dans notre globe

Les Eléments de la philosophie de Newton

Précis du siècle de Louis XV

Dictionnaire philosophique, éd. Naves et Benda, Garnier, Paris. 2 vols.

Lettres philosophiques, éd. Naves, Garnier, Paris.

Traité de métaphysique, crit. ed. H. Temple Patterson, Manchester University Press, 1937.

Wade, Ira O. *The Clandestine Organization and Diffusion of Philosophic Ideas in France from 1700 to 1750*, Princeton, 1938.

Whiston, William *A New Theory of the Earth, from its Original, to the Consummation of all Things*, 3rd ed., London, 1724. (1st ed. 1696)

Woodward, John *An Essay towards a Natural History of the Earth, and Terrestrial bodyes, especially Minerals*, 3rd ed., London, 1723. (1st ed. 1695)

INDEX

GPSR Authorized Representative: Easy Access System Europe - Mustamäe tee 50, 10621 Tallinn, Estonia, gpsr.requests@easproject.com

www.ingramcontent.com/pod-product-compliance
Ingram Content Group UK Ltd.
Pitfield, Milton Keynes, MK11 3LW, UK
UKHW021658040826
473956UK00004B/342
9780691627144